What Comes after Entanglement?

a Cultural Politics book

Edited by John Armitage,

Ryan Bishop,

and Douglas Kellner

WHAT COMES AFTER ENTANGLEMENT?

Activism,
Anthropocentrism,
and an Ethics
of Exclusion

EVA HAIFA GIRAUD

Duke University Press *Durham and London* 2019

Printed in the United States of America
on acid-free paper ∞
Cover designed by Courtney Leigh Baker
Text designed by Aimee C. Harrison
Typeset in DIN Neuzeit Grotesk and Quadraat
by Copperline Books

Cataloging in Publication Data is available
from the Library of Congress.

ISBN 9781478005483 (hardcover : alk. paper)
ISBN 9781478006251 (pbk. : alk. paper)
ISBN 9781478007159 (ebook)

Cover art: *Ciclotrama 16, Pequena história sobre o pecado* (A short story about sin), 2014.

For Greg

Contents

Acknowledgments ix

Introduction 1

1 Articulations 21

2 Uneven Burdens of Risk 46

3 Performing Responsibility 69

4 Hierarchies of Care 98

5 Charismatic Suffering 118

6 Ambivalent Popularity 142

Conclusion: An Ethics of Exclusion 171

Notes 183

Bibliography 225

Index 241

Acknowledgments

In particular, I want to say thank you to Greg Hollin, Tracey Potts, and Isla Forsyth; the ideas that are central to this book were informed by our collaborative work. The relationship between entanglement and exclusion was something that emerged as particularly important though these collaborations, and the term *ethics of exclusion* that I engage with throughout the book is taken from our piece "(Dis)entangling Barad: Materialisms and Ethics," in *Social Studies of Science*. Other ideas central to the book have emerged out of cowritten research with Greg alone, particularly the discussions of care ethics and laboratory beagles. These collaborations have been really important to me, on both a conceptual and a personal level: it just always feels like such a pleasure and privilege to work together.

Several other friends and colleagues have been incredibly generous with their time and volunteered to read various sections of this book, as well as other related articles. Special thanks to Andy Balmer, Josh Bowsher, Des Fitzgerald, Marie Thompson, and Greg, again, for offering to do this. Any errors and issues, of course, are my responsibility alone.

The advice I have received, at all stages of the review process, from colleagues at Duke has been invaluable. I can't express how useful the advice from my editor, Courtney Berger, has been in particular and would also like to thank Jenny Tan for helping me gather together materials in the final stages. Perhaps above all, the incisive and constructive suggestions from the two anonymous reviewers have played an integral role in informing the final version of the book, and I really want to express my gratitude. It's difficult to articulate how valuable I found the advice I was given.

I would also like to thank the editors of the *Cultural Politics Book* series. Particular thanks to John Armitage and Ryan Bishop for their support with my proposal. Thanks, too, to Mark Featherstone for his early encouragement. I also need to extend thanks to those who supported me at the very first stages of this project. The initial idea emerged during my PhD at the Centre for Critical Theory at the University of Nottingham, so thanks to my supervisors Tracey Potts (whom, as I mention above, I've been lucky to continue working with since) and Neal Curtis; my internal examiner, Colin Wright; and my external examiner, Jenny Pickerill. In addition, this book could not have been written without my experiences working with Veggies Catering Campaign, which offered insights that not only informed my conceptual arguments but really brought home the importance of everyday difficulties that have to be navigated in activist practice. Thanks in particular to Patrick Smith and Brent Reid.

Over the past few years I've had the opportunity to present or discuss work at various conferences and workshops, where I've had some really helpful, critical feedback; these include the Social Movements and Media Technologies seminar series, Coproduction of Knowledge with Nonhumans stream (at the 2014 Royal Geographical Society annual conference), and the workshops Working across Species, Digital Food Activism, and Digital Food Cultures. I've also been lucky to organize streams and workshops with some fantastic people, such as Media Environments (which I coconvened with Neil Archer and Pawas Bisht) and the 2017 Science in Public stream Animals in Public (coconvened with Angela Cassidy). During this period I have been especially appreciative of encouragement from people at the more "critical" edge of the animal studies spectrum, but who also engage with relational theoretical work. Particular thanks go to Seán McCorry as well as colleagues involved with the Lund University Critical Animal Studies Network, especially Tobias Linné, who has generously invited me to teach sessions in his module Animals in Society, Culture and the Media at Lund for the past three years. It is always such an inspiring experience working with everyone involved with the network.

More broadly, thank you to friends and colleagues at the Centre for Critical Theory and Institute for Science and Society at Nottingham, and the people I work with in my current role at Keele. Thanks in particular to the colleagues who have shown me real generosity during the writing process, whether this has been through offering friendship, mentorship, and support or just offering an encouraging comment when I needed it: Michelle Bastian, David Bell, Jen Birks, Anita and Roger Bromley, Ian Brookes, Susan Bruce, Carlos Cuevas

Garcia, Caroline Edwards, Anna Feigenbaum, Maria Flood, Annette Foster, Andy Goffey, Eleanor Hadley Kershaw, Richard Helliwell, Katie Higgins, Wahida Khandker, Lydia Martens, James Mansell, Ivan Marković, Deirdre McKay, Robert McKay, Ceri Morgan, Warren Pearce, Stefanie Petschick, Lynne Pettinger, Laura Maya Phillips, Ed de Quincey, Sujatha Raman, Barbara Ribeiro, Kathryn Telling, Teodora Todorova, and Tom Tyler.

Particular thanks go to my colleagues in the media department at Keele, whom I feel really lucky to work with: Pawas Bisht, Sam Galantini, Mandy McAteer, James McAteer, and Vicki Norman, and—of course—to the students I've had the privilege of teaching at both undergraduate and postgraduate level. I'm particularly appreciative of working with some fantastic PhD students at Keele. Special thanks are due to Elizabeth Poole and Wallis Seaton, whose work has been important in enabling me to rethink my own approach.

Finally, thank you to all of my friends and family, and especially Annie Giraud, Abdulrahman Giraud, and Danny Giraud. I love you.

Introduction

Some things are impossible to disentangle. It has become increasingly commonplace to argue, for instance, that humans are never autonomous beings who act against an essentialized natural world; instead, the human is only realized by and through its relations with other entities.[1] Animals, of course, play a significant role in constituting what it means to be human—inhabiting positions that range from valued domestic companions and livestock to nuisance "trash animals" or uncharismatic invasive species—but so too do technologies, microbes, and minerals.[2] The labels that are used to designate other creatures and materials betray further complications, by pointing to the role of all manner of taxonomies, values, cultural associations, and practices in shaping how particular human communities relate to other beings.[3]

Yet, although some entanglements might be too messy to unpick, they have also offered a source of ethical and political potential.[4] By foregrounding the ways that human existence is bound together with the lives of other entities, contemporary cultural theorists have sought to move beyond a worldview where the human is seen as exceptional. Narratives of entanglement have, in such contexts, proven important in implicating human activities in ecologically damaging situations and calling for more responsible relations to be forged with other species, environments, and communities.[5]

Actually meeting these responsibilities, however, is not a straightforward task. Irreducibly complex situations—where human and animal lives, ecological processes, and technical arrangements are impossible to meaningfully separate—cannot be settled by neat solutions that focus on one factor alone. From this perspective, issues such as seaborne plastic pollution cannot be solved by placing the blame on poor waste disposal practices on the part of certain communities, as this masks the role of particular relations of production, leaky plastic recycling chains, and the material properties of plastic itself in constituting the crisis.[6] Likewise, singling out specific agricultural chemicals as to blame for declines in biodiversity fails to recognize how agricultural practices are bound up with commercial infrastructures, animated by more-than-human agency, and imbricated in geopolitical inequalities.[7] These are just two examples that speak to a broader theoretical emphasis on the need to avoid imposing simplistic solutions on difficult, multifaceted problems—solutions that not only fail to do justice to these problems but can actively cause damage in their moralism.[8]

While emphasizing the complex, more-than-human entanglements that constitute lived reality has proven politically and ethically important, such an approach also carries dangers. Though it might be important to recognize the nuances of a given situation, this can also make it difficult to determine where culpability for particular situations really lie, let alone offer a sense of how to meet any ethical responsibilities emerging from these situations.[9] Irreducible complexity, in other words, can prove paralyzing and disperse responsibilities in ways that undermine scope for political action.[10]

My aim in posing the question "What comes after entanglement?" is not to deny the entangled complexity of the world, therefore, but to explore the possibilities for action amid and despite this complexity. Throughout the book I elucidate a number of tensions that have emerged in relation to existing attempts to ground an ethics and politics in the recognition of relationality. These tensions have not been caused by the act of acknowledging the complex, coconstitutive relations that tie diverse actors together, but have been generated by the assumption that more ethical—or at least less anthropocentric—modes of action necessarily follow from this recognition. The phrase "what comes after," then, is not just intended to underline the need to develop a fuller account of the types of ethics that can emerge from relationality, but to pose deeper-rooted questions about the value of a relational emphasis for grounding ethico-political practice.

In response to this line of questioning, I argue that in order to create space for intervention, there is a need for a conceptual reorientation.[11] Rather than focus on an ethics based on relationality and entanglement, it is important to more fully flesh out an ethics of exclusion, which pays attention to the entities, practices, and ways of being that are *foreclosed* when other entangled realities are materialized.[12] By developing this argument throughout the book, I elucidate that although narratives of entanglement grasp something important about the world, they do not capture everything. Attention also needs to be paid to the frictions, foreclosures, and exclusions that play a constitutive role in the composition of lived reality. Centralizing and politicizing these exclusions, I contend, is vital in carving out space for intervention.

As a whole the book is concerned with this broader conceptual attempt to engage with exclusion as a means of grappling with the paradoxes of relationality and attendant difficulties associated with action and intervention. To underpin this overarching theoretical focus, however, I draw particular inspiration from activist practice. Each chapter engages with interrelated instances of anticapitalist, animal, and environmental activism, moving broadly chronologically from the 1980s to the present day: from anticapitalist pamphleteering campaigns, activist experiments with digital media, and food activism in protest camps, to controversies surrounding laboratory animals and popular environmentalisms on-screen. Some of the practical difficulties faced by particular groups, related to resourcing, discursive constraints, or infrastructural limitations, might appear mundane. I argue, though, that even the most everyday problems hold significant theoretical implications by elucidating the frictions—and even dangers—in realizing relational modes of ethics within concrete contexts of political contestation. The specific difficulties faced by the groups I engage with are not just informative in themselves, therefore, but speak back in productive ways to work that has grappled with the political and ethical implications of living in entangled worlds. This work has been undertaken across the humanities and social sciences, including in science and technology studies, animal studies, the environmental humanities, more-than-human geographies, and bodies of cultural theory such as new materialism and posthumanism.

As well as foregrounding the difficulty of acting amid complexity, the activist groups to whom I turn highlight how paying more conceptual attention to exclusion can provide a route beyond these difficulties. Any given sociotechnical arrangement—from a fast-food restaurant to a media technology—materializes a particular way of doing things and creates norms and standards. If these norms are taken up on a large scale, they can easily become *normative*, presented as an inevitable or even natural way of organizing everyday life. As Susan Leigh Star argues, the congealment of infrastructural norms has ethical ramifications for those who do not fit with, or those who are excluded by, the systems at stake.[13] These arrangements, moreover, are often entangled with ways of thinking and acting that naturalize them and that themselves become difficult to challenge. The forms of environmental, anticapitalist, and animal activism I discuss throughout the book offer conceptual inspiration through the ways they engage with precisely this task of making particular norms visible in order to denaturalize them, ask questions about who or what is excluded, and—more important—find ways of contesting these exclusions (often by presenting alternatives).

Exclusions, however, are not just created by systems and institutions in ways that foster marginalization or oppression. As I illustrate throughout this book, it is important to recognize that all epistemologies or political and ethical approaches—even complex, pluralistic, and seemingly open ones—carry their own omissions. Any attempt to highlight or oppose systems that are perceived to be oppressive necessarily creates exclusions of its own, as it is sometimes necessary to contest certain relations in order to clear space for alternatives (indeed, this is often central to feminist and antiracist struggle). In such contexts, therefore, particular forms of exclusion, refusal, and opposition play a productive and creative, rather than wholly negative, role.

The inevitability, and indeed constitutive role, of some form of exclusion in any situation or environment means that it is neither something that can be avoided nor something that is intrinsically negative. What I argue throughout this book is that it is nonetheless important to make exclusions visible, in order to foster meaningful forms of responsibility for and obligation toward them. The problem, in conceptual terms, is that it is precisely this task of making exclusions visible that is difficult to realize if the conceptual emphasis is placed on relationality and coming together. An emphasis on the entangled relations that compose a given situation is not enough to bring the equally critical exclusions that are forged by it into view. This emphasis can also obscure who bears the greatest burden of these relations. Centralizing exclusions, in contrast, holds potential for opening them to future contestation and the possibility of alternatives that could better spread these burdens. As I elucidate throughout the book, therefore, emphasizing and politicizing exclusion is not just a means of complicating narratives of entanglement but offers alternative trajectories for grounding ethical and political intervention.

Before I develop these arguments, it is thus useful to gain a clearer sense of why it is so urgent to find an alternative means of supporting political and ethical action, once the entangled composition of the world has been acknowledged.

Entanglement

The fragility of particular forms of life and ways of living has been brought increasingly to the fore, not just in the biological and earth sciences but across work within the social sciences and humanities.[14] In response to growing concerns about anthropogenic problems—from mass extinction and climate change to more everyday but equally contentious issues surrounding everyday practices in farms and laboratories—the past three decades have seen the emergence of new interdisciplinary fields. The rise of animal studies, the environmental humanities, and extinction studies, for instance, has resulted in difficult questions about human obligations being brought to the fore. Beyond early transhumanist and posthumanist interventions, a burgeoning body of work has emphasized the ways that human existence has *always* been knotted together with the lives of other entities.[15] The purpose of emphasizing these histories of entanglement is to move beyond discourses of human exceptionalism, which can be used to justify practices that are damaging to those deemed nonhuman, other-than-human, or less-than-human.

Although the significance of more-than-human agencies has long been recognized in certain strands of science studies and geography, within the growing fields of animal studies and the environmental humanities such understandings are increasingly positioned less as making conceptual or ethical claims about reality and more as offering a simple recognition of the way things are.[16] The manner in which the more-than-human has been figured in theoretical contexts, then, has evolved conceptually, with a gradual shift from narratives of hybridity to assertions of entanglement.[17]

Hybrid figures and environments, from genetically modified mice to cityscapes, have been central to critical-feminist theory since the 1980s and played an important role in challenging the notion of the epistemic purity of categories (chief among them nature and culture).[18] Entanglement furthers this line of argument by encapsulating the myriad of world-making relationships that constitute environments, relationships that are irreducible because they are not just interactions between discrete actors that can be disentangled with the right conceptual or indeed political tools.[19] These relations are, instead, the site through which subject (and object) positions, identities, and even materialities themselves emerge. Undergirding these developments, therefore, is a departure from the sort of epistemological concerns central to Bruno Latour's *We Have Never Been Modern*, toward Donna Haraway's ontological assertion that "we have never been human."[20]

By engaging with these developments, I am not seeking to homogenize a heterogeneous body of work, which has evolved in different ways in different disciplinary contexts, but rather to explore the far-reaching implications of tendencies that are shared across these approaches. These tendencies, broadly speaking, involve decentering the human as the locus for ethics and politics through recognizing—and often celebrating—relationality.[21] The act of foregrounding the entangled composition of particular environments has, in turn, led to a rejection of totalizing ethical frameworks that are insufficiently responsive to this dynamism and complexity.

Attempts to move beyond the human are thus bound up with broader attempts to move beyond the *humanist* epistemologies that are seen to support anthropocentrism. Nonanthropocentric perspectives have worked to respond to the complex set of environmental problems that—it has been alleged—are underpinned by liberal-humanist modes of relating to the world.[22] The use of humans as a benchmark for all ethical concerns is seen to have had catastrophic consequences, because this exceptionalist logic ensures that, no matter how messy ethical decisions are, as long as they benefit humans in the last instance, then the problems caused for nonhumans are a necessary (if sad) sacrifice.[23] This is not to say that from a humanist perspective the problems facing nonhumans do not matter, are not seen as damaging, or are seen as not having significant consequences, but that the logic of human exceptionalism ensures that human benefit is the ultimate arbiter.

For example, from a relational, more-than-human perspective, it is human exceptionalism that inhibits restrictions on emissions, reductions in consumption, or further regulation of human engagements with animals within the agricultural-industrial complex. Humanist commitments ensure that issues, conversely, are made to *matter* politically only to the extent that they impact humans; in line with this perspective, for instance, climate change is seen as warranting action only if it affects people and perhaps even then only certain types of people (with economic benefit often triumphing over environmental concerns).

To combat these problems, there has been a push to unsettle anthropocentric humanism in favor of a more relational understanding of the world that recognizes and engages with more-than-human agencies. Indeed, some of the most urgent conceptual work has aspired to make critical interventions beyond theoretical debates, in settings ranging from conservation and environmental activism to neuroscience, pedagogy, fine art, and quantum theory.[24] This work has proven critically important in conceptualizing the entangled composition of lived reality and the dynamics of knowledge production. The problem, I suggest, lies in work that has made assumptions about the types of ethics and politics that follow from these entangled onto-epistemologies. In general terms there is still a tendency to celebrate entanglement—or treat it as a good in itself—with questions about intervention hinted at but ultimately left underdeveloped.[25] Yet simply acknowledging that human and more-than-human worlds are entangled is not enough in itself to respond to problems born of anthropogenic activity. As Alexis Shotwell argues in her otherwise-sympathetic engagement with relational ethical perspectives: "The specifics of how we would understand and act on the specifically ethical call [these bodies of work] make are somewhat thin. In these texts, theorists do not tell us how to parse the specifics of the ethical call, or the relational economy toward which we might aim to behave more adequately."[26] It is dangerous to assume, therefore, that less anthropocentric forms of ethics and politics automatically proceed from the recognition of relationality, at least not in a straightforward sense. The problem is that relational approaches do not just make intervention difficult but actively problematize conventional modes of ethics and politics because relationality—as a conceptual commitment—is, in part, constituted by a resistance to ethico-politics that is perceived to lack this complexity. The paradox of relationality, in other words, is that it struggles to accommodate things that are resistant to being in relation, including forms of politics that actively oppose particular relations.

Obligation and Responsibility

Action and intervention are especially hard to accommodate within relational, more-than-human theoretical work because of the way that resistance to anthropocentrism is bound up with a broader wariness of humanism. From this theoretical perspective, commonplace political frameworks for extending questions of justice beyond the human are inadequate. The extension of rights to animals and environmental actors is treated with suspicion because such a stance mirrors an exceptionalist logic that shores up human privilege.[27] This line of argument is typified by Haraway's claim that "we do not get very far with the categories generally used by animal rights discourses, in which animals end up as permanent dependents ('lesser humans'), utterly natural ('nonhuman'), or exactly the same ('humans in fur suits')."[28] This argument speaks to a broader point about the devastating consequences of species hierarchies that can arise when, for instance, certain charismatic megafauna (usually those attractive to humans) are afforded protection while other animals and ecosystems remain "killable."[29] Exceptionalism can also reinscribe colonizing and indeed colonialist logics, as with clumsy attempts by large nongovernmental organizations to raise awareness of practices such as seal clubbing or dog meat production, without attending to the way these tactics can be damaging for Indigenous communities or perpetuate ethnocentric stereotypes.[30]

The other reason why narratives of rights have been problematized, which has been stressed in the posthumanities in particular, is that such expansions can lead down a conceptual rabbit hole where the central preoccupation is who (or what) gets to count as having rights once the concept is applied to animals. Do invasive species count? Do mosquitos? What about deadly viruses?[31] Although questions of "where rights end" might seem facetious, they point to important concerns about the dangers (as Jamie Lorimer puts it) of grounding "appeals for animal rights on the comparable existence of essential human characteristics in non-humans" and thus only "extending the franchise to certain privileged others."[32]

Where relational approaches to ethics have been critically important is in drawing attention to some of the tensions associated with frameworks such as rights, in ways that hold implications for particular instances of activism and advocacy.[33] What becomes concerning is when these arguments move beyond the contestation of *specific* modes of advocacy or argumentation, to become a more broad-brush condemnation of so-called totalizing critique, as crystallized by Latour's infamous "critique of critique" or illustrated by the splintering off of critical scholarship within particular fields such as animal studies.[34] Though these developments have led to some productive academic and activist trajectories, the treatment of work labeled "critical" as somehow marginal and lacking nuance has had worrying consequences, especially when it comes to addressing questions of action and intervention.

As I illustrate throughout the book, both certain strands of academic work and particular modes of political intervention are routinely sidelined for being overly critical. To revisit the example of critical animal studies (CAS), for instance, although certain strands of CAS display a blanket suspicion of theory, other work with critical commitments has engaged more sympathetically with relational ethics.[35] Yet the important conceptual interventions that have been made by this body of critical work are often not taken seriously within "mainstream" animal studies or allied fields; instead work that critiques contemporary human-animal relations is routinely portrayed as stemming from a naive commitment to totalizing frameworks (and dismissed on this basis).[36] This body of scholarship, in other words, is portrayed as failing to do justice to the nuance of multispecies entanglements, which means the specific *content* of critical arguments can then be neglected.

As I argue within the main body of this book, CAS is just one example of critical, oppositional thought and practice being marginalized because of its lack of fit with relational modes of ethics. These forms of marginalization are concerning as they can inadvertently reinforce existing social inequalities that make it difficult for certain communities to articulate a critique of particular social norms.[37] What is especially dangerous about the marginalization of these forms of critique is that it marks a failure to do justice to the epistemological and ethical work that overtly critical perspectives—perhaps even those incommensurable with relational approaches—can accomplish.

These dangers are elucidated by Maria Puig de la Bellacasa, who foregrounds how theoretical work that insists on complexity, nuance, and "a balanced articulation of the involved concerns" is often incompatible with the sort of critical perspective that can "produce divergences and oppositional knowledges based on attachments to particular visions, and indeed that sometimes presents its positions as non-negotiable."[38] For Puig de la Bellacasa, the omission of critical perspectives is worrying because "these are voices required to support a feminist vision of care that engages with persistent forms of exclusion, power and domination in science and technology," and this potential can be shut down if only perspectives that are articulated in nonanthropocentric language are engaged with.[39]

The apparent incompatibility of particular ethical stances and forms of political intervention with relational approaches, therefore, actively places epistemological limits on theoretical work that seeks to move beyond the human. In addition, foreclosing dialogue with critical work also has stark ethical implications, and it is these implications that become apparent when shifting the focus away from what sort of ethics can emerge from the entanglement in itself, to instead flesh out an ethics of exclusion.

The Work of Exclusion

A small, but critically important, interdisciplinary body of scholarship—spanning more-than-human geographies, the environmental humanities, and science studies—has called for greater recognition of the undesirable nature of certain forms of relation and the need (in certain contexts) to preserve distance, alterity, and separateness.[40] Rosemary-Claire Collard's important research on the global wildlife trade, for instance, elucidates the dangers of predicating an ethics on entanglements and encounters between species. As Collard argues: "An essential part of forming animals' commodity lives in global live wildlife trade is that their wild lives are 'taken apart' in that they are disentangled from their previous behaviors and ecological, familial, and social networks."[41] For particular, commodified, relations to emerge, others have to be undone. These processes of commodification, moreover, can never be fully unpicked, with wildlife rehabilitation centers struggling to reentangle primates with the relations they were removed from. For Collard, notions of naturecultures give little sense of how to distinguish between these two very different "relational economies" (to revisit Shotwell's turn of phrase). While remaining suspicious of essentializing notions of nature, therefore, Collard suggests that some notion of wildness might nonetheless be worth recuperating in order to oppose particular relations that, once accomplished, cause harm or violence that can never be completely reversed.

Tensions associated with relational ethics are not just evidenced by dramatic examples such as the global wildlife trade; Franklin Ginn, for instance, elucidates how even everyday activities such as gardening result in all manner of encounters where distance and exclusion offer more ethical purchase than being in relation. Ginn's theoretical intervention "Sticky Lives" engages with the slug as a figure whose relations with gardeners are fraught with ethical difficulties, due to the incommensurability of particular forms of relation. Simply put, plants, gardeners, and slugs cannot thrive in the same place at the same time, so the act of tending to a garden necessarily involves decisions about how to manage these slimy gastropods. Despite the damage slugs wreak, Ginn found that gardeners were reluctant to kill them outright and engaged instead in all manner of experiments to create a slug-free space: from throwing them over fences for birds to eat, to creating physical barriers, or even cultivating herbaceous borders entirely from plants disliked by slugs. Ethical connection with slugs, in other words, was negotiated not through attachment but through finding alternative ways to *detach* slugs from gardens. The desire for nonrelation, or, as Ginn beautifully phrases it, the way that gardeners "create spaces around hoped-for-absence rather than relation," elucidates the inevitability of exclusion, then, but also its ethical potential.[42]

Though the global wildlife trade and everyday practices in gardens might be very different examples, they both point to particular forms of work that can be accomplished by exclusions. For Collard, refusing or opposing particular relations (through recuperating some form of alterity and wildness) is necessary in order to preserve others, while for Ginn exclusions play a constitutive role in creating the garden as a space where certain plants (and indeed those who tend to them) can flourish. These sorts of examples do not just point to the need, in certain instances, for distance or disentanglement, then, but foreground that the act of excluding certain relations is precisely what creates room for others to emerge, or for existing forms of life to be sustained. Exclusion can also, therefore, be a site where accountability is taken not just for who or what is classified as an actor worthy of moral consideration, but—more fundamentally—for which worlds are materialized over others.[43]

Building on these arguments, I suggest there is a need to recognize not just the constitutive role of certain exclusions but their productive role: that purposeful acts of contesting particular relationships are sometimes necessary to create space for alternatives to emerge. It is important, however, not just to recognize the role of exclusion but to foreground its ethical and political significance. It is in addressing questions of the ethical and political *work* that certain exclusions accomplish that informative lessons can be learned from activist practice, as elucidated through turning to some commonplace issues within women's and anticapitalist movements. These tensions are encapsulated by Jo Freeman's classic text, "The Tyranny of Structurelessness," which illustrates a tension at the heart of movements that aspire to reject social hierarchies. Her focus is on groups that are organized in a structureless, leaderless way and in which decisions are reached by consensus and everyone—ostensibly—has the right to speak. Freeman points out, however, that structurelessness brings its own tyrannies. The problem is that "contrary to what we would like to believe, there is no such thing as a 'structureless' group. Any group of people of whatever nature coming together for any length of time, for any purpose, will inevitably structure itself in some fashion."[44] Unlike the hierarchies Freeman's women's groups are working to contest, these emergent structures are informal and born, for instance, of friendships that might exist externally to the group, of the confidence or rhetorical abilities of particular group members, or even of technical skills that particular members have (or are perceived to have). For all their informality, these hierarchies have very concrete consequences and inform how roles are distributed within groups, whose voices are heard the most clearly, and whose ideas ultimately inform practice. Inevitably, these relations tend to be imbricated in classed, raced, and ableist inequalities. Making space *available* for people to speak in the group itself, therefore, is not enough, as relations that existed before or outside of the group situation continue to foreclose possibilities for certain participants to take up these opportunities, while making it far easier for others. These barriers, crucially, cannot necessarily be seen when focusing on encounters or relations within the group itself but only become visible on tracing longer intersectional histories.

What is especially pernicious about informal hierarchies is that because they occur in an ostensibly structureless or nonhierarchical space, the persistence of inequalities is often rendered invisible. The disproportionate influence of certain people's opinions, or heightened perceptions of their abilities, can thus be naturalized (with particular individuals perceived as having the best ideas, or being best suited to a particular role). What is still more problematic is that within a nonhierarchical situation these informal hierarchies cannot be challenged, as no one has the authority to do so, and any such attempt is perceived as reinserting or imposing authority in a space that explicitly opposes such expressions of power. These problems are intensely difficult to negotiate and have resonance beyond women's groups, with Freeman's arguments regularly drawn on to account for informal hierarchies that emerge in contexts from digitally mediated activism to university classrooms.[45]

Yet, just as the problems associated with informal hierarchies have persisted in contexts beyond activism, Freeman's tactics for navigating them also have wider purchase; she argues, for instance, that certain structures are necessary, but only if they are designed to distribute power evenly and make it visible. Any rules should ensure that "the group of people in positions of authority will be diffuse, flexible, open and temporary" so that privileged individuals "will not be in such an easy position to institutionalize their power."[46] To be structureless in any meaningful sense, for Freeman, then, requires a degree of structure to ensure accountability and responsibility.

Although Freeman's conception of informal hierarchies has had a profound legacy within social movements, I argue that her arguments also hold conceptual significance in the context of relational, more-than-human theoretical work. What Freeman's work foregrounds is that in order to create alternative ways of being, it is necessary to make decisions not only about which relations to prefigure and enact but about which to exclude. These decisions, however, need to be temporary, contingent, and open to contestation to ensure they do not congeal in ways that allow normative social relations to simply reimpose themselves and reinscribe existing inequalities. If these arguments are related back to the theoretical contexts at stake here, this points to a particular conceptual problem: as Ginn puts it, an emphasis on entanglement and relationality can "ignore the non-relational, what may not be vital, and what may precede or be obscured by existing relations."[47] At present, I argue, the exclusions fostered by relational theoretical work itself are insufficiently visible, because attempts to engage in such criticisms are frequently—to use Puig de la Bellacasa's evocative turn of phrase—thrown out "with the corrosive bathwater of critique."[48] If these tensions in relational, more-than-human perspectives are read against Freeman, what emerges as important, then, is recognizing that purposive decisions to exclude certain relations do not have to be negative, and are indeed inevitable, but that it is nonetheless critically important to find clearer ways of fostering responsibility for these exclusions. It is in realizing ways of taking responsibility that especially important lessons can be learned from the instances of activism discussed within this book.

Finding Affinities (and Frictions)

As hinted at by the insights that can be gained from Freeman, throughout the book I tease out some of the ways that activist work can offer insight into how to act in contexts that are resolutely complex, by revealing barriers in translating theory into practice and tactics for negotiating these barriers. There are, therefore, numerous reasons for finding affinities between particular strands of theory and practice, but one particular factor makes achieving this dialogue both especially helpful and especially difficult: the way that very different perspectives share a superficially similar vocabulary. The language of openness, riskiness, experimentation, and ecology is used by some of the social movements I draw on and by social movement theories, as well as relational, more-than-human approaches. As I make explicit, however, it would be a mistake to assume this shared terminology equates to shared meaning, and rather than neatly mapping theory onto practice (or vice versa), it is necessary to adopt a more diffractive approach.

Karen Barad, following Haraway, advocates an approach that moves beyond reflexive approaches to cultural theory in favor of diffractive ones. To elucidate what a diffractive methodology entails, she describes the process of two stones being dropped into water. Each stone creates ripples, but as they come together, a more complex diffraction pattern emerges as the two sets of ripples converge and complicate one another. By attending to the pattern that emerges as the ripples meet, Barad suggests, it is possible to learn something of the apparatus that produced it. This diffractive methodology offers a means of "reading insights through one another in attending to and responding to the details and specificities of relations of difference and how they matter."[49]

The process of understanding how subtly different perspectives and practices can occupy shared spaces and complicate one another captures something of the messy relationships between the strands of theory and practice I discuss throughout this book. The activist initiatives discussed here often adopt communicative tactics, or share vocabulary and values, which seem to have a sympathetic relationship with theoretical work. For instance, as I elucidate throughout the book, particular activist communities appear to share the theoretical concern with complexity, storytelling, openness, nonhuman agency, care, and affect. In practice, however, the tactics used by activists often cut against the types of politics and ethics called for in theoretical contexts, due to being grounded in normative appeals to social justice or questions of suffering. In focusing on these tactics, then, I aim to attend to the sort of specific differences that Barad describes in order to explore how and why these emergent tensions matter in conceptual as well as practical terms.

The tactics I draw on as a lens through which to diffract tensions between theory and practice are derived from an interrelated range of initiatives, where activists have sought to communicate their arguments to wider publics. Beginning with anticapitalist fast-food activism, which originated in the 1980s (in chapter 1), I then move on to early activist experiments with digital media (chapter 2), performative activism within protest camps and free-food giveaways (chapter 3), tactical attempts to contest mainstream media discourses about antivivisection activism (chapter 4), and social media campaigns surrounding laboratory beagles (chapter 5). The book culminates with a focus on popular media where arguments articulated by early grassroots movements seem to have gained mainstream attention via Hollywood-backed features and globally marketed documentaries (chapter 6).

The instances of activism engaged with in each chapter offer privileged sites for drawing out tensions associated with core theoretical debates. The first chapter, for instance, traces affinities and tensions between work in feminist science studies that has emphasized relationality and entanglement and tactics engaged in by activists locked in a court battle with the fast-food corporation McDonald's. Bringing these perspectives into conversation highlights some of the core difficulties in articulating issues without reducing their complexity or smoothing out their messiness. Indeed, what is argued in this chapter is that insisting on a particular model of articulation (that takes relationality as its baseline) can sometimes make it difficult for particular communities to speak at all. This problem is picked up in the second and third chapters, which focus on groups who have sought to actively transform the infrastructures that make complex articulations and interventions difficult, by creating alternative arrangements to provide food, communication networks, and even sewerage systems. These examples appear to embody cosmopolitical modes of risky, experimental politics. They also, however, foreground dangers that can arise when risks are not spread evenly, due to being distributed in ways that reinscribe gendered, geopolitical, and racial inequalities.[50]

The second half of the book develops these arguments further and explores particular tensions that have emerged in relation to the politics of care and emotion, first as knowledge politics, then in relation to somatic ethics, and, finally, in the context of affective media imagery.[51] The fourth chapter turns to controversial campaigns surrounding primate research and situates them in relation to speculative care ethics, to highlight the ways in which particular theoretical arguments can inadvertently foster hierarchies of care that delegitimize the emotional and affective work engaged in by activists. The final chapters then tease out the stakes of these hierarchies of care by tracing how particular tactics (such as an emphasis on suffering) and emotional registers (such as uses of sentimentality) are positioned negatively in relation to embodied modes of care and affect that have been advocated in theoretical contexts. While I aim to recuperate these concepts, my aim is not to do so uncritically but to simply pay greater attention to the work that they achieve in order to explore the potential for pushing them in less anthropocentric directions.

The Personal and the Political

In addition to being an especially useful site for making the ethical stakes of intervention and exclusion visible, the movements focused on here have been selected, in part, due to my own engagement with animal activism. Particular groups have been focused on due to issues that emerged through my own participatory action research with grassroots food activists, which led to me either working with or becoming aware of the campaign tactics engaged in by affinity groups working on different issues. My work with these groups necessitated an understanding of the longer histories of anticapitalist fast-food campaigning, which are the focus of the first chapter. I also had to make extensive use of the activist media technologies that are foregrounded in the second chapter (as well as becoming acutely aware of the strengths and shortcomings of these media). My discussion of food and media activism in chapters 2 and 3 draws extensively on my experiences facilitating free-food giveaways, and the tensions we had to negotiate in realizing these actions, while particular affective experiences inform the final chapters.

As will be elaborated on throughout the book, I am drawing attention to my involvement with some of the groups I discuss here for three key conceptual reasons. First, it is politically and ethically important to situate this book. Although particular experiences help to anchor some of the theoretical arguments made throughout each chapter, this approach necessarily has its limitations, and a notable issue here is that I predominantly focus on groups working in the Global North (although this is not universally the case). It is thus important to situate my arguments in this context. Despite the limitations of a situated approach, it is important in refusing a universalizing stance, even as I work to cautiously tease out the more profound provocations particular groups offer to contemporary theoretical work.

The other two reasons for adopting a situated approach pertain to the way that my own experiences triggered particular conceptual questions, which motivated me to write this book. Even though the groups I focus on are working within very particular contexts, they still unsettle certain ways that intervention and action are currently conceived in theoretical contexts. My own personal experiences of food activism have provided insight into the differences between, on the one hand, the ways activists are represented (in both mass media and theoretical contexts) and, on the other hand, the mundane practicalities of activism, particularly when it comes to the task of self-representation. Activism is messy, and activists are often constrained by particular legal systems, media narratives, and communications infrastructures, to name just a few commonplace barriers. When one is engaged in campaigning work, it is easy to resort to emotive imagery and abstract slogans, just to make some sort of a difference while working within and against the constraints of a given system. The pragmatic compromises resorted to by protest movements are precisely what has led to these groups being criticized within contemporary cultural theory for promoting overly simplistic solutions to irreducibly complex problems. These everyday constraints, therefore, are not just practical problems but conceptual ones, in actively inhibiting the aspirations to practice that are hinted at by the theoretical work at stake here.

The final reason why I have drawn attention to my own experiences pertains to the relationship between affect, emotion, and praxis. The value of recognizing the role of affect, and creating space for emotion in a more sustained way, has been seen as pivotal to activist practice, especially in relation to public engagement; there has, correspondingly, been a push to legitimize affect and emotion, in order to counter technocratic discourses that portray activist perspectives as irrational due to their emotional commitments.[52] In theoretical terms there has been a parallel move to foreground the importance of emotion, with recent conceptual work aiming to legitimize emotional responses but deconstruct the dichotomy between "rational" science and "emotional" publics by foregrounding the (often-valuable) role of emotion in scientific practice and transspecies communication.[53]

There are again, however, tensions between theory and practice regarding how different forms of affective or emotional encounter are depicted and understood. In theoretical contexts the mundane affects that are generated as experts or specialized workers interact with animals in their everyday caretaking and conservation work are often portrayed as holding ethico-political potential, in giving rise to sustained relations of care.[54] In contrast, activist emotions are often portrayed as lying at the root of paternalistic or irrational responses to political issues, or even as giving rise to problematic forms of anthropomorphism wherein human emotions are attributed to animals.[55] Often the specific emotional responses activists describe when viewing certain images, or engaging in certain practices, are precisely what is sidelined in theoretical texts, after being cast as sentimental and anthropomorphizing (even as the role of emotion is valorized in other contexts).[56]

Overall, therefore, it was often the disjunction between particular practical experiences, and emotions, on the one hand, and the ways these forms of activism were represented in theoretical contexts, on the other, that motivated me to explore the stakes of these tensions. In asking questions about these issues, I am not straightforwardly defending ethico-political frameworks that are routinely used in problematic ways, but working to create space for understanding specific instances of critical, oppositional, and activist thought in more ambivalent ways. Refusing to sideline "critical" perspectives out of hand means that the insights gained from them can be taken seriously in conceptual terms and offer productive ethical and epistemological provocations for contemporary theoretical work.

In general, therefore, while findings from participatory action research have informed my arguments about how tensions between theory and practice manifest themselves in activism, the purpose of this book is not to provide an ethnographic or auto-ethnographic account of specific movements. Instead, I aim to draw inspiration from situated practices in the work of particular groups and use them to flesh out informative tensions between theory and practice. Teasing out these relationships, in turn, helps to lay the groundwork for crafting an ethics grounded in the recognition of not entanglement but the constitutive and in some instances creative role of exclusion.[57]

A (Final) Note on Tactics

The word *tactic* is used throughout the book to characterize approaches that have proven valuable in navigating the core problems that each chapter focuses on. The term has been chosen deliberately, due to its connotations of contingency and resistance. Tactics, to echo Michel de Certeau, offer context-specific approaches for resisting power.[58] Though de Certeau is not drawn on in a sustained way here, as some sort of conceptual touchstone, the notion of tactics nonetheless captures something of the approaches outlined throughout the book, which do not offer a universalizing template for political action but emphasize context-specific praxis. At the same time, tactics are a useful concept in maintaining a focus on how praxis is framed by power. *Power* itself is a term that—like activist standpoints—has sometimes been ejected from relational theories due to being perceived as a totalizing explanatory framework that lacks context-specific nuance.[59] De Certeau, in contrast, offers an alternative trajectory for conceiving of power, where the term does not serve as a totalizing category but is indicative of attempts by certain actors to control others by imposing regulatory strategies on their movements and cultural practices. Urban planning, for him, is the archetypal example as roads, pavements, and barriers are all put in place to encourage certain movements and discourage others. Tactics (which could include everything from leaping over barriers, crossing the road somewhere other than a crossing, or engaging in more creative acts such as parkour) are the processes of resistance to these strategies, which reveal their fissures and points of weakness.[60] While for de Certeau *tactics* are all manner of everyday microsociological processes of resistance (conscious or not), I am using the term here in reference to more conscious and critical forms of activism.

Though the recurring argument made throughout the book is that activist practice is conceptually informative, the text is not designed to generate one-way traffic and treat practice as simply a tool that enhances theory. The hope is that theory can also help to foreground particularly valuable tactics for decentering the human, amid the myriad of approaches that constitute activist protest repertoires. This approach, however, demands a reevaluation of the notion of tactics itself; as discussed above, in de Certeau's use *tactics* refers to acts that take place within the confines of a system or territory and that are engaged in by those who do not govern this territory or set its rules. If understood in line with this specific understanding of *tactics*, therefore, the work of activists is seen as operating on a terrain in which their actions can navigate preexisting structures, and toy with these structures, but as lacking the capacity to reshape the territory itself. De Certeau's formulation is thus a useful starting point, but the approaches advocated in this book aim to go beyond simply acting within territory, in order to actively *intervene* in it. Though they do not change the overarching rules of the game wholesale, I nonetheless argue that the tactics explored throughout the book can make (and have made) interventions that prompt responses, instigate material-semiotic reconfigurations, and open possibilities for political change.

Haraway argues that "some actors, for example specific human ones, can try to reduce other actors to resources—to mere ground and matrix for their action. . . . [S]uch a move is contestable, not the necessary relation of 'human nature' to the rest of the world." Moreover, "other actors, human and nonhuman, regularly resist reductions. The powers of domination do fail sometimes in their projects to pin other actors down."[61] The tactics outlined throughout the book evoke different ways of approaching the project articulated by Haraway, offering different means through which activists and researchers can "increase the failure rates" of actors attempting to reduce others to "mere ground and matrix" for their action. Unlike strategies, tactics do not seek to impose their own way of doing things (and thus become activist norms in themselves) but suggest how context-specific and contingent practices could be used to contest the processes through which social actors—both human and nonhuman—are treated as resources.

The "tactical interventions" I foreground, therefore, are not intended to be prescriptive but are nonetheless valuable in drawing attention to and contesting different modes of conceptual and sociotechnical exclusion. This approach is important in light of the sympathetic critique of particular modes of more-than-human, relational ethics that underpins this book: the recognition that no form of relation is innocent is insufficient in accounting for the exclusions that are bound up with any form of relation. The need to take responsibility for exclusions, however, does not mean that they are a bad thing; as well as being constitutive, they can also be creative and ethically important. Certain exclusions, in certain situations, might be necessary in spreading the burden, resisting oppressions, and creating space for new ways of doing things to come into being. It is nonetheless vital to find far clearer ways of fostering obligations toward these exclusions. What I elucidate throughout the book is that the recognition of entanglement—in particular, the entanglement of humans and other actors—does not intrinsically create room for such obligations, or necessarily give rise to less anthropocentric ways of thinking and acting in the world. Indeed, in some instances affective relations and entanglements can be instrumentalized or can marginalize critical perspectives.

Perhaps, then, asking what sort of ethics and politics can emerge from entanglement is the wrong framing of the question. Although some things are impossible to disentangle, recognition of this complexity does not capture everything about material reality, and, as such, this emphasis does not offer as helpful a foundation for ethics and politics as it might seem. Instead, more concerted efforts need to be made to render visible—and assume ethical responsibility for—the exclusions that play an equally constitutive role in materializing particular realities at the expense of others.

1 | Articulations

The power of feminist analysis is to move from the experience of being a non-user, an outcast or a castaway, to the analysis of the fact of McDonald's (and by extension, many other technologies) and implicitly to the fact that "it might have been otherwise"—there is nothing necessary or inevitable about the presence of such franchises.
—Susan Leigh Star, "Power, Technology and the Phenomenology of Conventions"

There is a much more fundamental problem than Big Macs and French Fries: capitalism. . . . Alternative and radical ideas have spread throughout society, drawing on past experiences, on present situations, and on people's hopes and practical visions for the future. What are the global alternatives to a system based on profits and power? How can a society be created which is based on the principles of human solidarity and mutual aid, on sharing and co-operation, on freedom, and on harmony with the environment and respect for life? How will people be able to run such a society together? None of these questions are new—there is a wealth of ideas and experiences documented from the past or from more recent struggles and movements which people can learn about and take strength from. That is one of the purposes of McSpotlight. We must continue to develop the ideas and activities which are laying the basis of a new society within the shell of the old.
—"Capitalism," McSpotlight.com

Susan Leigh Star's characterization of feminist analysis does not retain its value purely because of its ongoing relevance to science and technology studies.[1] Beyond the field in which it originated, Star's call to explore how things "might have been otherwise" can be used to illuminate vital affinities—and equally vital tensions—between relational, more-than-human theoretical work and particular activist movements.

The above extract from the McSpotlight website, for instance, seems to share the central concern of feminist science studies in asking what alternatives (in this instance to global capitalism, as crystallized by McDonald's) might look like in practice. At the same time, a closer look at the protest ecologies underpinning this message illustrates some persistent difficulties associated with attempts to articulate alternative visions of the world. It is these everyday barriers to articulation, I suggest, that lie at the heart of tensions between political practice and contemporary theory that seeks to move beyond the human.

A number of challenges faced by anticapitalist activists are outlined throughout this chapter, ranging from resourcing issues to sociolegal arrangements and questions of public engagement. Such difficulties speak to a specific point: the setting in which political action unfolds can often undermine activists' attempts to convey the complex relationships among interrelated issues, here the intersection of environmental, labor, and animal welfare concerns.[2] These everyday challenges often place activist practice at odds with a theoretical commitment to an ethics grounded in the recognition of relationality and irreducible complexity. Seemingly mundane difficulties in activist organization, in other words, are not merely practical matters but hold conceptual significance.[3]

As I illustrate throughout the chapter, what makes things especially hard is that everyday challenges within activism can be actively exacerbated by the modes of cultural politics offered by theoretical attempts to respond to entangled worlds. Constraints on practice, for instance, often produce particular identity positions, tactics, and affective logics that are at odds with the tenets of theoretical work but are nonetheless valuable in making interventions.[4] In light of these constraints, an insistence on emphasizing complexity and relationality can—in certain contexts—make it difficult to speak at all.[5]

Issues that make it difficult to realize relational modes of ethics are often interpreted as problems that need to be negotiated in order to offer a clearer sense of how such approaches can emerge in practice. These tensions are, in other words, often framed as practicalities within activism, or problems generated by common modes of advocacy, which need to be overcome in order to realize more nuanced, multifaceted articulations of issues. This chapter, in contrast, argues that tensions between strands of theory and activist practice are not always things that can—or indeed should—be worked through in order to enact theoretical demands. Tensions between conceptual work and activist practice do not necessarily speak to problems within *practice* (or at least they do not just do this) but can be used to elucidate deep-rooted *theoretical* issues associated with relational, more-than-human modes of ethics.

Throughout this chapter I foreground informative points of tension between theory and practice, arguing that these tensions do not just indicate a need to complicate existing relational approaches but illustrate why an alternative ethical orientation is needed. In doing so, the chapter lays the foundation for the rest of the book and its exploration of how a focus on exclusion can help to provide this orientation.

Anti-McDonald's protest is a valuable site to turn to because of its capacity to foreground, in ways that speak to contemporary theoretical work, the difficulties that have recurred in subsequent large-scale instances of anticapitalist protest. Now over twenty years old, the McSpotlight website, for instance, was originally a hopeful symbol that represented the power of anticapitalist counternarratives to contest sociotechnical arrangements that fostered inequality.[6] Though these hopes have since waned with the decline of the global justice movement, anti-McDonald's campaigning, as a key forerunner of campaigns that linked anticapitalist concerns with environmental and animal activism, created important legacies for contemporary activism (to the point that it is now almost seen as an anticapitalist cliché!).[7] Important lessons can nonetheless be learned from revisiting early anti-McDonald's protest, not only due to its legacy for more contemporary movements but due to insights provided by its own rich history. In particular, the campaign speaks to difficulties that can arise when activists seek to move beyond "single-issue" politics to instead articulate the complex relationships among different issues.[8] It is this question of articulation that I focus on here, because it offers a productive means of grasping the stakes of the frictions that can exist between theoretical work and political practice, even that which seems to aspire to the same ends.[9]

McDonald's and McLibel

What became known as the McInformation Network was predated by anti-McDonald's campaigning that had been occurring on a small scale in the United Kingdom throughout the 1980s (notably, the first International Day of Action against McDonald's by the London Greenpeace campaign group was held on October 16, 1985).[10] The campaign emerged and grew exponentially after McDonald's attempted to sue two activists, Helen Steel and Dave Morris, for distributing a six-page fact sheet that was critical of the corporation, serving libel writs against them in 1990.[11] The fact sheet's original purpose was to draw together long-standing concerns about the corporation held by activist groups working in different contexts: criticisms ranging from environmental damage to animal welfare concerns, and from advertising targeted at children to workers' rights.

The company's decision to sue Steel and Morris (who were to become "the McLibel Two"), somewhat ironically, transformed the series of UK-based demonstrations into an international campaign. In addition to a McLibel Support Campaign being established to support the UK activists, a transnational McInformation Network was founded in order to gather and document further critical information about McDonald's (with the McSpotlight website ultimately serving as the hub for this material). The decision by McDonald's to undertake legal action, in other words, is what elevated the protest to the global stage befitting the issues it was addressing. The company's actions gave the campaign unprecedented levels of publicity, as the trial itself lasted almost three years (from June 28, 1994, to June 19, 1997) and the solidarity website McSpotlight was allegedly accessed 2.2 million times on the judgment day.[12] Indeed, when McSpotlight was launched in 1996 it received media attention in its own right and, as touched on in chapter 2, gained broader academic and activist attention in elucidating the internet's potential to support dissent.[13]

As conveyed by the McSpotlight quote that opened the chapter, productive dialogue can exist between elements of the campaign and theoretical work that has emphasized relationality as the foundation for situated ethics. The first half of the chapter, accordingly, focuses on tactics used during anti-McDonald's campaigning and the McLibel trial that resonate with theoretical work (particularly within feminist science studies and new materialisms). Anti-McDonald's activism is especially valuable in drawing out these affinities due to being contemporaneous with theoretical work that emerged in the early 1990s—including that of Star and Donna Haraway—that has gone on to shape the contours of debate both within feminist science studies and across related fields.

As argued by Star, when dealing with cultural phenomena such as McDonald's, feminist analyses go beyond mapping "the enrolment and *interessement* of eating patterns, franchise marketing, labour pool politics, standardization and its economics" to both ask who bears the brunt of these relations and contest their inevitability.[14] The story of anti-McDonald's campaigning has informative parallels with the mode of politics outlined by Star (and not just due to maintaining a common focus); the very purpose of these campaigns was to reveal what Star describes as "invisible work," the everyday exclusions and processes of marginalization that enable an actor like McDonald's to exist.[15]

Tactics engaged in by activists also have broader affinities with conceptual work, in seeming not only to offer potential to realize less hierarchical modes of campaigning but also to lend themselves to less anthropocentric praxis. To intervene in existing norms instantiated by McDonald's, for instance, it was necessary to denaturalize the infrastructural arrangements of McDonald's through articulating the liveliness and agency of actors—human and non-human—who were bound up with these arrangements. This approach often involved foregrounding frictions within the infrastructures of McDonald's itself, in order to denaturalize these arrangements and articulate the possibility of other forms of organization.[16] As I go on to outline in the first half of this chapter, the tactics of critical articulation that activists engaged in thus appear to segue with theoretical calls to create space to ask whether problematic norms "might be otherwise."

Yet, despite its success, the campaign also suffered difficulties, and it is these difficulties that offer points of tension with theoretical work that hold (perhaps unexpected) conceptual significance. As I go on to argue in the second half of the chapter, activists' opportunities for developing critical articulations of McDonald's were constrained by particular legal apparatuses and media arrangements, which made radical-participatory communicative tactics difficult to realize. Mundane barriers relating to financial restrictions, legal requirements, and the dynamics of activist media ecologies meant that it was often difficult to involve the actors who were most affected by McDonald's infrastructures in the work of articulating these relations. These restrictions also meant that it was difficult to articulate the complexity of the issues at stake in a way that met the requirements of legal evidence.

At first glance many of the obstacles that constrained participatory forms of protest appear to be decisively practical issues, related to money, resourcing, and technical problems, but these obstacles nonetheless carry conceptual freight due to lying at the root of tensions between critical-activist perspectives and theoretical work. The McLibel case and McSpotlight website are thus helpful in foregrounding issues that inhibit a wholesale departure from representational modes of advocacy in order to embrace more multilayered modes of articulation. These difficulties were made all the more profound when those being represented were not human (as with the environmental and animal welfare issues at the heart of the campaign). To better understand these tensions—and the broader theoretical provocations they offer—it is useful to first sketch out the modes of articulation that have been called for in theoretical contexts as a point of comparison.

Anthropocentrism, Representation, and Articulation

The tensions surrounding articulation that I focus on within this chapter relate to a broader theoretical concern with contesting paternalistic modes of advocacy that seek to speak for others. Thinkers such as Haraway have repeatedly drawn attention to the dangers of advocacy work that "pleads the cause of another" because it instills "a power relationship not unlike those of guardianship or parenthood."[17] This line of argument has proven not just long-standing but influential and deserves attention because it lays the groundwork for subsequent conceptual perspectives that have displayed even greater suspicion of certain forms of advocacy work.[18]

Though it comes to the fore in her work on companion species, Haraway's own wariness of representational advocacy is set out most forcefully in an earlier essay where she describes this approach as a "political semiotics of representation."[19] She characterizes this politics as a representational approach to advocacy that insists that it is necessary to speak for those who cannot speak for themselves. This aim is often thought of as innocuous, or even necessary in certain contexts, but can lead to inadvertent forms of political ventriloquism that reinscribe inequalities rather than overturning them.

Haraway's arguments are explicitly grounded in postcolonial commitments.[20] For instance, she illustrates the danger of representational forms of advocacy by drawing on a series of deeply problematic representations of the Amazon rain forest from the late 1980s and early 1990s to interrogate how certain campaigns depicted Indigenous Kayapó communities as vulnerable to industrial encroachment and in need of protection. Though some of these campaigns are now thirty years old, Haraway's critique is still relevant, as a similar logic has persisted in more recent initiatives. For instance, the importance of remaining suspicious of a political semiotics of representation is underlined by a more recent internet meme entitled "This Image Should Be Seen by the Whole World," which emerged in 2011 but continues to circulate today.[21] This brief, widely disseminated social media post contained a photograph of Chief Raoni Metuktire in tears, accompanied by text that read:

> While magazines and TV chains report about the lives and love affairs of movie actors and actresses, football players and other celebrities, the Chief of the Kayapo tribe heard the worst news of his entire life:
>
> Mrs. Dilma, the president of Brazil, has given her approval for the construction of an enormous hydroelectric central (the world's third largest one).
>
> This means the death sentence for ALL the tribes living at the shores of the river because the barrage will flood more or less 988,421 acres of the forest. More than 40 000 natives will have to find other living surroundings where they will be able to survive. The destruction of the natural habitat, the deforestation and the disappearance of several species of plants and animals will be a fait accompli.
>
> We know that a simple image is the equivalent of a thousand words, it shows the price to be paid for the "quality of life" of our so-called "modern comforts." There is no space in the world anymore for those who live differently. Everything has to be smoothed away, that everyone, in the name of globalization must lose his and her identity and way of living.
>
> If this enrages you, I urge and implore you to "SHARE" this message to all your friends, relatives and acquaintances.
>
> Thank you in the name of life, nature and biodiversity.[22]

This text, and its attendant imagery, crystallizes the central features of a political semiotics of representation. For Haraway, what lend support to problematic modes of representation, like those employed by the above post, are sharp bifurcations between ontological categories (such as nature versus culture) and the ethical and epistemological hierarchies these categories are entwined with. Here "life, nature and biodiversity" are constructed as needing to be spoken for, and it is presented as the responsibility of those in the Global North to speak up for the Kayapó (and communities in the Amazon region) in order to preserve their "way of living." The approach taken in the meme thus rushes to denunciate industrialization and offer straightforward solutions to the social and environmental problems it engenders but, in doing so, forces entangled concerns to fit neat ethical narratives.[23] In addition to adhering to a logic of representation that reproduces hierarchical orderings of relations, this form of representation undermines the work of those closest to the situation at stake.[24] Alongside the meme's paternalistic sentiment, for instance, the accompanying photograph echoes Haraway's wariness of the way images used in advocacy often strip those being depicted from their "constituting discursive and non-discursive nexuses," enabling them to be "relocated in the authorial domain of the representative."[25]

Simply put, therefore, the meme offers an illustration of how a political semiotics of representation privileges advocates by situating them within the domain of modernity, while those being represented are positioned within the realm of untouched nature (as illustrated here by the stark dichotomy the meme draws between "modern comforts" and the lives of Indigenous communities). As Haraway argues, moreover, such paternalistic modes of advocacy are not just a rhetorical misstep but lend support to structures that consecrate the difficulty of particular actors speaking for themselves, through creating relations that ensure that "tutelage will be eternal. The represented is reduced to the permanent status of the recipient of action, never to be a co-actor in an articulated practice among unlike, but joined, social partners."[26]

Although the contestation of nature/culture dichotomies is now a well-trodden path, the existence of this contemporary meme illustrates the continued relevance of Haraway's early criticisms of advocacy that is predicated on these distinctions. As exemplified by national-park models of conservation, dichotomies between nature and culture not only persist within dominant modes of advocacy but underpin environmental initiatives ranging from the biopolitics of conservation to popular awareness-raising films.[27]

Exchanges surrounding "This Image Should Be Seen by the Whole World" are useful for grasping the stakes not only of Haraway's critique of representation but the alternative approach she offers, "a political semiotics of articulation."[28] While the original post circulated widely, it was also heavily criticized. For instance, a response blog post emerged shortly after the original meme and was itself circulated extensively. However, unlike the original, the second blog quotes Chief Raoni directly, who condemns the original post:

> I did not cry because of the authorization for construction and the beginning of the work of Belo Monte. As long as I will live, I will continue to fight against this construction. I want to tell President Dilma, to Lula, to the President of FUNAI, to the President of IBAMA, to the Minister of Energy Lobão, that I am on my way to Brasilia and that I will take along all my warriors to fight against the Belo Monte. I will not stop.
>
> It is President Dilma who will cry, not me. I wish to know who published this picture and spread this information. I would like to see this person.[29]

The post also foregrounded the work that Kayapó activists had themselves engaged in to challenge the dam—campaigning, protesting, and forging a global solidarity network— and pointed out that this ongoing work needed to be centralized and supported, rather than displaced by ethnocentric advocacy narratives. Follow-up reflections from the blog's author provided the rationale behind their initial critique, making explicit that their response was an attempt to combat representations of "Indigenous people as powerless and in need of non-Indigenous benevolence to survive . . . that allows us to see a picture of a crying man and use it to justify silencing his voice."[30] What these exchanges underline, then, is the way that representational modes of advocacy can actively reproduce the cultural relations they are (ostensibly) attempting to contest, as well as the need for alternative modes of articulation.

Like the approach taken by this second post, what Haraway pushes for in a political semiotics of articulation is a refusal to speak *for* those perceived to be exploited, to instead articulate *with* those who are affected by the issues at stake. The countermeme illustrates potential ways of achieving such articulations that resonate with Haraway's arguments, as it emphasizes the need to create platforms for the voices of those affected by a particular concern (here by engaging with Chief Raoni's own representation of the issue) rather than (mis-)representing them in line with a predefined ethical agenda. As discussed in this and subsequent chapters, a range of sustained and complex alternatives to representation can also be found in the participatory forms of research engaged in within fields such as social movement studies and radical geography, as well as within social movements themselves (all of which create tensions as well as carrying potentials). Here, however, I dwell on a different aspect of Haraway's approach to articulation, her attempt to extend questions of "articulating with" (rather than speaking for) beyond the human.[31]

Due to understanding distinctions between nature and culture as being entangled with other inequalities, a political semiotics of articulation has—at its foundation—an emphasis on articulating the more-than-human composition of environments. This approach is underpinned by the assertion that the recognition of multispecies entanglements is a prerequisite for overcoming the colonizing consequences of humanist logic.[32] What this form of politics entails on the ground might seem a little unclear in comparison to representational modes of advocacy, but Haraway illustrates what a more productive approach could look like by drawing on Susanna Hecht and Alexander Cockburn's *The Fate of the Forest*.[33] She argues that their work helps to deconstruct "the image of the tropical rainforest . . . as 'Eden under glass,' which needs to be separated and protected from man," and praises their work for instead telling "a relentless story of a 'social nature' over many hundreds of years, at every turn co-inhabited and co-constituted by humans, land and other organisms. For example the diversity and patterns of tree species in the forest cannot be explained without the deliberate, long-term practices of the Kayapó and other groups."[34] Hecht and Cockburn's arguments, in other words, are praised by Haraway for their characterization of the forest as a space constituted by relations among its human and nonhuman inhabitants (a concept that has since gained currency with concepts such as naturecultures). "The Promises of Monsters" thus foreshadows both Haraway's more recent work (such as her narratives of sympoietic relationships between species) and the broader centrality of relationality within the environmental humanities, while also offering a distinct sense of the importance of relationality for nonhierarchical political *practice*.[35] To reiterate Haraway's argument as to the political value of this form of articulation: "Some actors, for example specific human ones, can try to reduce other actors to resources—to mere ground and matrix for their action; but such a move is contestable, not the necessary relation of 'human nature,' to the rest of the world. Other actors, human and nonhuman, regularly resist reductions. The powers of domination do fail sometimes in their projects to pin other actors down; people can work to enhance the relevant failure rates."[36] For Haraway, in other words, the articulation of relationality opens up ethical opportunities by challenging attempts by certain (privileged human) actors to reduce the capacities of others. Humanist arguments, for instance, such as those criticized by animal ethicist William Lynn—which frame nonhuman actors as "resources that [lie] beyond the boundaries of moral community"—are untenable for relational ontologies, because the human cannot be separated out as a distinct category, worthy of special treatment, in order to justify exploitative relationships with other entities.[37]

What is important about this line of argument is that Haraway situates her critique not in relation to humanity in a general sense but in relation to a specific, liberal-humanist conception of the human. This situatedness is critically important, ensuring that an emphasis on the relational composition of the world lends itself to a postcolonial as well as a nonanthropocentric project.[38] Resonating with Star, Haraway's approach turns attention to the infrastructural and material-semiotic relations that reproduce and naturalize inequality. In the process, Haraway foregrounds the ways in which anthropocentric orderings of relations can saturate socioeconomic and geopolitical inequalities by legitimizing the silencing or exploitation of those perceived as closest to the "natural world."[39]

In line with Star and Haraway, then, the aim of articulating the complex material-semiotic relations that lie behind particular modes of representation, or infrastructural arrangements, is to disrupt the inevitability of these relations in order to open space to ask whether things could be otherwise.

Contesting Norms within the McLibel Trial

Against the backdrop of these theoretical debates, the problematic implications of representational rights language are stark; this approach has been accused by Haraway and others of perpetuating asymmetries within advocacy work by reinforcing dualistic modes of thought (which work to marginalize particular actors) and action (which legitimate everything from political ventriloquism via social media to neocolonial conservation initiatives). Suspicion of representation is thus not just an abstract concern but entangled with very material stakes.

These stakes, for instance, are evident in early anti-McDonald's campaigning, especially the 1985 fact sheet that was central to the McLibel trial. When discussing global inequalities, the pamphlet explicitly criticizes depictions of those suffering due to structural inequalities as being helpless, arguing that commonplace narratives used within charitable campaigns "to get 'compassion money'" ultimately divert "attention from one cause: exploitation by multinationals like McDonald's."[40] This line of argument, however, is compromised by their broader narrative about environmental destruction (such as their critique of cattle grazing and attendant colonial relations in the Amazon rain forest) being framed instead in relation to the sort of "Eden under glass" imagery that Haraway criticizes: "AROUND the Equator there is a lush green belt of incredibly beautiful tropical forest, untouched by human development for one hundred million years, supporting about half of all Earth's life-forms, including some 30,000 plant species, and producing a [m]ajor part of the planet's crucial supply of oxygen."[41] While designed to convey what is in danger of being lost due to industrial agriculture, appeals to this imagery also undercut the arguments put forward in the rest of the pamphlet, which reveals a constant struggle to depart from reductive modes of representation. What reading Haraway's arguments against the fact sheet works to do, therefore, is foreground how a political semiotics of representation—predicated on sharp distinctions between humans and the natural world—is dangerous not just in itself but also in the ways it undermines attempts to critique geopolitical inequalities by shoring up the logic that underpins these asymmetries.

On first impression, what seems to be needed is thus to find some other way of translating the sort of relational ethico-epistemologies pushed for by Haraway into practice, in order to offer more complex, multifaceted articulations of specific issues that refuse a reductive logic of representation. What I argue in the remainder of the chapter is that things are not quite this straightforward and that while it is important to recognize helpful affinities between theoretical work and particular instances of activism, it is not always possible—or desirable—to overcome points of tension. To bring these problems into relief, however, it is useful to first draw attention to some productive commonalities between these instances of activism and the theoretical work at stake here.

Across the different media platforms used, and within the McLibel trial itself, what emerges is an approach that aligns more productively with Star's characterization of the project of feminist science studies. Instead of a straightforward advocacy campaign that seeks to speak for all of those who were somehow affected by McDonald's, the trial offered a space to ask questions about how particular cultural realities came into being and were sustained in order to ask whether things could be otherwise. More important, there was a concerted effort to involve those closest to these infrastructures in the articulation of the issues at stake. Central to the activists' defense was building up a picture of how infrastructures that were associated with McDonald's worked to enroll certain entities as consumers and others as consumed, or, to put things differently, how certain relations of capital were reproduced by these infrastructures, which positioned workers, farmers, and environmental actors in particular ways. For the McLibel activists, developing an understanding of the preexisting articulations of issues, and offering their own alternative articulations, offered a means of denaturalizing social norms and opened questions about social and ethical responsibilities for these norms.

One of the key issues that had to be overcome during the McLibel trial is encapsulated by Stengers and Philippe Pignarre, who characterize the refrain of capitalism as "sorry, but we have to."[42] For Stengers this logic is crystallized by the way the undesirable side effects of growth are depicted as an unfortunate consequence of a system to which there is no alternative, a system of "generalized competition, a war of all against all, wherein everyone, individual, enterprise, nation, region of the world, has to accept the sacrifices necessary to have the right to survive (to the detriment of their competitors), and obeys the only system 'proven to work.'"[43] What the McLibel case helps to elucidate is how such assertions are not simply rhetoric but lent ontological weight by the sort of "irreversible" infrastructural norms foregrounded by Star, norms that make alternatives not only difficult to envisage but almost physically impossible to enact. In the McLibel trial, for example, McDonald's repeatedly justified its treatment of animals, workers, and consumers through an appeal to "industry standards" while eliding its role in setting these standards, a strategy that posed specific problems for the activists. During the trial the activists had to provide conclusive evidence that McDonald's had done everything they alleged within the pamphlet, because the UK libel system places the burden of proof on the defendants.[44] McDonald's, in contrast, did not have to provide evidence to prove it was guilt free, which meant its defense could rest entirely on appeals to existing sociotechnical norms. McDonald's, in other words, could portray itself as following standards and even laws that were predetermined by infrastructural requirements; the corresponding construction of the social was, in turn, framed as the natural consequence of technical necessity. The tasks faced by the activists and McDonald's were thus deeply asymmetrical, not only because of the economic and legal expertise the latter had at its disposal, or even the burden of proof expected from each party, but also because of their contrasting relationships with existing social norms, which were underpinned by apparently stable ontologies.

What is significant about the McLibel trial is that the tactic used by McDonald's, of appealing to these unspoken social norms and preexisting industry standards, was not always successful, with Steel and Morris often succeeding in exposing these norms to ethical scrutiny. Crucial to these tentative successes was a form of ontological contestation of the sort characterized by Steve Woolgar and Javier Lezaun in their analysis of the role of everyday objects in reinforcing mundane regimes of governance. An object as everyday as a garbage bag, they argue, can be mobilized in ways that demarcate sharp distinctions between rational and irrational actors (in their example, put-upon publics versus overly bureaucratic councils who fine people for using the wrong type of bag).[45]

Within the McLibel trial, specific pieces of kitchen equipment played a similarly divisive role to Woolgar and Lezaun's garbage bag but instead served as sites where specific sets of labor relations were enacted; industrial fryers became delineators of the difference between skilled and unskilled labor, and Big Macs separated corporate actors who offered consumer choice from activists who (seemingly) wanted to restrict it. In the labor section of the trial, for example, the arguments used by McDonald's were contingent on pieces of equipment being seen as neutral objects that fulfilled appropriate roles in food production, as this underpinned the argument that workers' wages were not low (as alleged by activists) because people were simply paid for the "jobs they do."[46] To contest this sort of "commonsense" assertion, activists had to articulate a very different fryer, one that systematically deskilled laborers. Pivotal to their arguments was unpicking the effort it had taken (on the part of McDonald's) to establish fryers as neutral objects, through tracing the cultural histories and infrastructural role of this piece of kitchen equipment in mediating labor relations. Burgers, similarly, were enacted by the McDonald's legal team as one possible food choice among others, in order to detach them from the systems that lay behind these choices. In contrast, activists had to bring these systems into view to draw out the ethical implications of particular ways of producing food.

As Woolgar and Lezaun point out, the stabilization of even the most everyday objects as normal is not a given but requires work; this point is ethically significant because if "entities are not given, but rather offer a reference point for temporary imputations of moral orders of accountability," then this opens space to show how "it could be otherwise."[47] Contestations about kitchen equipment fulfilled this role within the trial, from discussions of how breakdowns resulted in fat clogging drains, to revelations about floor-cleaning equipment not being used in practice due to (as one witness stated) it being "cheaper and quicker to use a mop and bucket."[48] In the final verdict, the success of these tactics was borne out as, although the activists did not win on all points, their approach had a degree of success: it was ruled that McDonald's was not simply paying people "for the work they do" but had developed sociotechnical arrangements that actively "depressed wages in the catering industry as a whole."[49] Similar ontological contestations occurred in relation to other entities that became focal points in the trial. McDonald's burgers, for example, shifted from being a neutral food option to something dangerously unhealthy, while the company's advertising was seen as exploiting children even though it was common practice in the advertising industry as a whole.[50]

The activists' contestation of norms was not just oriented around destabilizing the objects that gave ontological weight to the narrative of "sorry, but we have to," but extended to a broader contestation of the way in which these objects worked together to (re-)produce particular social realities. In the animal welfare portion of the trial, for example, suddenly things that were seemingly irreversible norms within farming practices became open to debate and discussion and even the acknowledgment that existing laws were inadequate in terms of making ethical judgments as to the well-being of animals.[51] As

David Wolfson argues, during the trial the "classic position of agribusiness" was treated as the benchmark for acceptable behavior by McDonald's itself, a position that at the time was the "Customary Approach" in gauging whether a practice was lawful, even though this meant that "any practice in accordance with common modern farming or slaughter practices [was] acceptable to the law, even if it [was] cruel."[52] During the trial, however, Chief Justice Bell "unequivocally rejected the Customary Approach stating he could not accept it for use in the case. He correctly noted that 'to do so would be to hand the decision as to what is cruel to the food industry completely, moved as it must be by economic as well as animal welfare considerations.'"[53] Under the criteria used to judge the libel case at least, this meant that the McLibel Two were able to challenge the use of particular infrastructural arrangements to naturalize and legitimate specific ethico-legal values. In other words, the activists had disentangled what was legally acceptable (i.e., anything that was a standard part of intensive farming practices) from the normative ethical world these sociotechnical arrangements brought into being. As a result of these tactics, for instance, the verdict troubled the unproblematic construction of male chicks as killable without ethical reflection, and posed broader questions about the wholesale exclusion of chickens from federal welfare laws. In doing so, the trial reopened debate about "facts" that were actually the product of infrastructural standards, to ask whether these infrastructures could be configured differently, with almost the entire animal cruelty portion of the trial being won by the defendants.[54]

Ontologically Disruptive Tactics: Invisible Work and Infrastructural Frictions

The point Haraway makes in "The Promises of Monsters" is that it is not enough to contest norms, or ask whether things could be otherwise, but that it is also vital to achieve this in a way that avoids displacing the voices of those most affected by these norms. While a political semiotics of articulation stresses the value of articulating with other actors, instead of speaking for them, the McLibel trial (and its attendant protest campaigns) reveals that this process is far from straightforward when the actors concerned are more-than-human, especially when physical distance exists between activists and particular objects of concern.[55]

It is in relation to this question of how to articulate with nonhuman actors at a distance that the tactic of foregrounding the "invisible work" of actors enrolled in particular infrastructures assumes particular resonance. *Invisible work* is meant here in two senses; it refers, first, to the labor that is involved in constructing and perpetuating sociotechnical networks, which is often hidden. As Maria Puig de la Bellacasa highlights, this work is routinely undervalued due to social relations that trivialize or render invisible the work that is involved in making these infrastructures function (especially forms of work that are classed, gendered, and racialized).[56] This type of invisible work, for instance, could include the unacknowledged affective labor demanded of McDonald's workers, such as the need to smile (an issue foregrounded by the initial anti-McDonald's fact sheet). The labor of the farmers, kitchen workers, and factory workers who produce happy-meal toys could, similarly, be seen as invisible work, especially in light of longer histories of the systematic removal of animal markets from public space and the globalization of food production.[57]

The second, related, understanding of *invisible work* put forward by Star relates to her assertion that "the public stability of a standardized network often involves the private suffering of those who are not standard."[58] In Star's terms, this type of invisible work is undertaken by everyone who does not fit the standards prescribed by these infrastructures, and who has to work harder to compensate for this lack of fit. Star uses her own onion allergy, an anomaly that McDonald's cannot accommodate in its standardized systems, to illustrate this form of work: "My small pains with onions are on a continuum with the much more serious and total suffering of someone in a wheelchair barred from activity, or those whose bodies in other ways are 'non-standard.' And the work I do: of surveillance, of scraping off the onions, if not of organizing non onion-eaters, is all prior to giving voice to the experience of the encounters. How much more difficult for those encounters which carry heavier moral freight?"[59] It was this form of work that the initial anti-McDonald's pamphlets were trying to make visible in their discussion of kitchen practices and farming processes, and although the pamphlet itself slipped into representational discourse at times, more complex articulations of invisible work unfolded in their other communicative practices.

The McLibel trial itself (and its online documentation) provided three years' worth of evidence to illustrate points where fast-food infrastructures had broken down; these breakdowns invariably occurred when the actors that McDonald's was attempting to enroll offered unexpected points of friction. In the workers' rights section of the trial, for instance, the ideal of perfect worker flexibility was shattered by evidence from employees who felt exhausted and undermined by their working conditions. Crucially, even nonhuman entities emerged as unruly forces when unexpected entities disrupted the restaurants' uniform production of food: from bacteria (which allegedly caused food poisoning) to congealed burger fat that blocked drains and led to sewerage flooding the kitchens of one particular restaurant.[60] These moments were foregrounded during the trial in order to unsettle the relationship between the social norms appealed to by McDonald's and the sociotechnical arrangements that sustained these norms. Indeed, the justifications that McDonald's gave for its policies provided some of the most humorous (and hence the most publicized) moments in the trial. For example, in its attempts to counter the activists' argument that McDonald's had promoted unhealthy food to children under the guise of it being nutritious, the company explained this was due to its distinct, literal definition of *nutritious* as "containing nutrients" and that this meant Coca-Cola could be seen as nutritious (due to containing sugar for energy).[61] Aside from offering amusement, what these exchanges illustrate is the political potential of highlighting infrastructural friction where, in Anna Tsing's terms, universalizing norms rub uncomfortably against local conditions.[62] As McLibel foregrounds, once these tensions are exposed, it becomes difficult to renaturalize them and justify business as usual.[63]

Through the trial, in other words, a range of actors whose work was ordinarily rendered invisible were afforded a platform to participate in the complex articulation of McDonald's infrastructures instigated by activists. The approach taken in the trial (and attendant campaigns), in other words, resonated with Haraway's call to contest particular actors being reduced to "mere ground and matrix" for the agency of others. As described above, two tactics emerged as especially helpful for engaging in this form of politics: highlighting the invisible work that goes into supporting sociotechnical infrastructure (ideally by articulating with those who undertake this work) and foregrounding frictions between the local actors whom McDonald's attempted to enroll and the infrastructural standards that it set. These tactics might seem to offer only micropolitical intervention but had significant legal consequences and were important in illustrating where McDonald's technologies failed to translate ideals of uniformity and control.

It is also through these tactics that productive points of theoretical affinity seem to emerge. Activist tactics during the trial offered forms of ontological contestation that resonate with not only earlier work in feminist science studies but more recent discussions of the politics of new materialism. Stacy Alaimo, for example, makes a similar point in relation to contemporary environmental activism, suggesting that particular groups often focus on the liveliness of objects (when drawing attention to the toxic effects of chemicals or slow death caused by seaborne plastic) and that these tactics open promising affinities between new materialisms and environmentalism. Echoing the tactics engaged in by anti-McDonald's activists, for Alaimo environmentalists' frequent focus on objects shows that in order to raise concern about particular issues, it is often essential to establish the liveliness of matter. In relation to plastic activism, for instance, she argues that animating "plastic stuff not only underscores how harmful—if not malevolent—plastic can be, it struggles to convey a sense of the material agency that will prove plastic is doing harm, as this is necessary in being taken seriously and prompting political action."[64]

The McLibel trial thus seems to not only illustrate how insights from feminist science studies and new materialism *could* be translated into concrete instances of political contestation but also elucidate how such tactics are often *already* a routine component of a politics that seeks to unsettle cultural norms associated with particular systems.[65] It is for this reason that the forms of ontological contestation engaged in by activists could potentially offer a site of dialogue with theoretical work.

Yet at the same time as affinities between strands of theory and practice are recognized, it is also vital to address points of tension that cannot be easily dismissed. For all the potential of these tactics to carve out a space for more meaningful dialogue between theory and practice, more needs to be said about the *difficulty* of engaging in an ontological politics that depends on activist definitions of objects being accepted.[66] Affinities between theory and practice are not always quite as neat, or indeed successful, as the instances of plastic activism Alaimo outlines. Victories in relation to animal welfare and workers' rights, for instance, need to be contextualized in relation to other, perhaps more profound difficulties faced by activists in other sections of the trial. Despite the partial successes of the McLibel trial, the case also offers insight into not only the issues that inhibit activist attempts to emphasize relationality and liveliness but also the reasons that this approach might not always be the best course of action. In doing so, the trial helps to shed light on important points of theoretical concern that need to be addressed.

Due to its length, to an extent the trial was able to give a platform to the people who were the most implicated in McDonald's infrastructures—particularly workers—to offer their own testimony of events. The tactic of highlighting frictions offers a sense of how the work of more-than-human actors can likewise be incorporated into the articulation of multifaceted issues. It is important, however, not to overstate the power of these tactics or their affinity with theoretical work. During the trial and attendant campaigns, there were still limitations regarding the forms of articulation that were possible for activists, which stemmed from financial constraints due to their lack of legal aid.[67] The notoriously asymmetrical levels of funds available to McDonald's and the defendants, respectively, resulted in a stark disparity in who was enabled to speak and who was excluded from consideration. McDonald's was alleged to have spent £10 million over the course of the trial and used these funds to mobilize a large number of experts from a range of global contexts to give evidence.[68] Activists, in contrast, could only bring witnesses to the court using the approximately £40,000 they had earned through fund-raising.

The courtroom itself, therefore, crystallized the economic disparity the activists were attempting to critique, in ways that pose profound problems for attempts to engage in ontological contestation or more complicated articulations of the issues at stake. It was no surprise that the strongest portion of the trial (in terms of the amount of evidence provided, as reflected by the verdict) was the workers' rights section, which could draw on evidence directly provided by McDonald's employees who could afford to travel to the hearing.[69] Franny Armstrong's *McLibel* documentary poignantly underlines this point, in a scene where a young ex-McDonald's worker from Canada describes spending her own money to travel to the United Kingdom to provide testimony, due to her sense of injustice at the working conditions she had faced.[70] Self-funding in this way was not possible for everyone implicated in the case, and, in contrast, the section of the trial that the judge, Chief Justice Bell, found to lack sufficient evidence on the part of the defendants was the link between McDonald's cattle ranches and deforestation in the Amazon rain forest. Due to a lack of resources to transport witnesses from the areas in question, only fifteen witnesses provided evidence for the defendants. Still more problematically, when the activists *were* able to introduce testimony from expert witnesses, the evidence they provided was deemed insufficient.

It is when turning to these witness statements that some of the most profound tensions emerge in conceptual terms. The star witness in this section was the very Susanna Hecht whose work Haraway praises in "The Promises of Monsters," and here again she painted a nuanced picture of the forest as a coconstructed space that was being damaged not by "human" encroachments into untouched "nature" but by the "radically altered patterns of land distribution" brought about by attempts (on the part of multinational-sponsored cattle ranches) to exclude certain actors from that space and turn others into resources.[71] Yet—to draw on language from Hecht's witness description—her "20 years of research fieldwork in Amazonia," and resulting account of the "biotic and social consequences" of the practices of McDonald's, did not meet the requirements of the court, and the defendants entirely lost this section of the trial.[72]

The trial, therefore, reveals several difficulties in realizing a political semiotics of articulation in practice. The structures of the court, first, worked to perpetuate broader socioeconomic inequalities fostered by infrastructures associated with McDonald's, as well as the broader media and legislative apparatuses that these infrastructures intersected with, by making it difficult for the activists to themselves articulate the complex relations they wanted to unpick, let alone involve others in these articulations. In addition to perpetuating structural inequalities between activists and McDonald's, in relation to their respective capacities to speak, the trial reinforced a further set of inequalities between those who were able to participate in the case (predominantly people from the Global North) and those who were excluded (people from the specific regions that produced raw materials for McDonald's). Finally, and perhaps most significantly in conceptual terms, these structures excluded certain practices of articulation. Though Hecht's testimony attempted to foreground the agency of the diverse sets of actors implicated in the production of raw materials, in the court case the limitations of this approach were brought into painful focus.

The asymmetrical tasks of sustaining and contesting norms, when the former is lent support by long-standing infrastructural arrangements, raises the question of why activists should complicate this process further by engaging in complex modes of nonrepresentational (let alone nonanthropocentric) politics. As Cary Wolfe himself acknowledges—even immediately after advocating a posthumanism that moves beyond rights-oriented approaches—discourses of rights and normative frameworks for social justice could provide a more straightforward conceptual shortcut for engaging with the public and are perhaps necessary in certain contexts.[73]

Yet, although the McLibel case elucidates barriers to realizing more participatory and less anthropocentric approaches, it also underlines the importance of overcoming these barriers. The trial, for example, foregrounds how public engagement is an insufficient rationale for utilizing representational shortcuts (i.e., through simplifying issues or temporarily speaking for others), because representational rights language also undercuts activist aims in a more practical sense. Narratives about defending rights have a long-standing association with the same discourses of possessive individualism that corporations appeal to when depicting themselves as offering consumer choice. Throughout the trial, rights claims grounded in liberal individualism were constantly drawn on by McDonald's to make their practices seem innocent. Kenneth Miles, then Director General and Chief Executive of the Incorporated Society of British Advertisers, for instance, argued, "In the United Kingdom, as in most countries, advertising is seen as a legitimate part of commercial activity and is valued by customers. It reinforces the role of competition in ensuring high quality goods and services for the public, who understand that most advertising is designed to keep them informed about one competitive brand or service in competition with another. . . . I see the advertisements for McDonald's Restaurants as playing a construction [*sic*] part in showing both parents and children an additional and competitive choice in meal time opportunities."[74] As Annemarie Mol suggests, these sorts of strategies are designed to "shift the site of the decision elsewhere. . . . [T]hey displace the decisive moment to places where, seen from here, it seems no decision but a fact."[75] For Mol, the relations that emerge from particular assemblages are significant in that they do not mark just the production of a given reality but the *exclusion* of alternative ways of doing things.[76] For activists, therefore, avoiding the rhetoric of rights grounded in liberal individualism was critically important as this logic necessarily foreclosed the alternative visions of the world that they were committed to. Rejecting liberal-individualist notions of rights, in other words, is important in order to avoid reinforcing the logic that enabled McDonald's to present existing norms as both a technical necessity and a matter that was beyond ethical debate, in a manner that made alternatives not only difficult to realize but impossible to even conceive.

From Mundane Problems to Theoretical Dilemmas

Even though the difficulties faced by activists during the McLibel trial appear to be everyday problems, then, they pose conceptually significant questions. On one level, these questions speak to existing concerns about what sort of politics or ethics can emerge from the recognition of entanglement and complexity. An emphasis on relationality means that sometimes it is difficult to develop concrete responses to especially damaging forms of relation, due to there being no clear right answer.

This line of argument is encapsulated by Haraway's own calls to "stay with the trouble," which suggest that the courses of action offered by her own and related approaches are not necessarily easy but vital nonetheless. From this perspective, the irreducible complexity of the world is not something that it is possible to tidy away for the sake of developing more straightforward ethical solutions. If the tensions described throughout this chapter are read in line with these arguments, therefore, they could simply be read as markers of the type of trouble that concerns Haraway: wherein difficult, multifaceted material realities demand equally difficult and complex ethico-political responses.

As touched on in the introductory chapter, however, this theoretical emphasis has often resulted in suspicion toward perspectives that come from a social justice or animal activist agenda, due to the perception that these perspectives maintain totalizing viewpoints. Puig de la Bellacasa's account of "angry environmentalists" speaks to this point. In her own distinct articulation of care ethics, she draws attention to the epistemic and ethical inequalities that can inadvertently be reinscribed if one insists on particular models of relationality as a baseline. To reiterate Puig de la Bellacasa: "Respect for concerns and the call for care" that emanate from theoretical contexts can "become arguments to moderate a critical standpoint" and create "an obligation for the (environmental) activist to replace excessive critique and the suspicion of sociopolitical interests with a balanced articulation of the involved concerns."[77]

It is in relation to Puig de la Bellacasa's arguments that the everyday difficulties of realizing an ethics grounded in entanglement and complexity can be interpreted not as being signs of productive trouble but as posing more intractable problems that actively undermine certain forms of intervention. These problems are especially profound for those whose perspectives diverge from existing ways of doing things. As illustrated by McLibel, the structures of courts, or structural inequalities created by corporate infrastructures, often prevent activists from articulating the relational composition of a given issue.

In the trial, for instance, while the activists did succeed in contesting certain standards set by McDonald's—relating to animal welfare, the directing of advertising toward children, and workers' rights, for instance—some of their other arguments (such as their engagement with complex Amazonian ecologies) were seen as insufficient to support their legal claims, in ways that spoke to and compounded racial and geopolitical inequalities. It is important, therefore, to recognize the difficulties posed when activists have to work in situations that not only are governed by the norms that they are opposed to but lend ontological support to these norms. To go back to Puig de la Bellacasa's point, it is dangerous to always insist on a "balanced articulation" of the issue at stake, not only because this requires the modulation of critical voices but because—in some contexts—the resources required to articulate the complexity of an issue are unavailable to those whose perspectives are already marginalized. Insisting on a particular mode of political articulation can, therefore, inadvertently foreclose alternative perspectives while leaving the status quo untouched.

The difficulties facing activists also point to further tensions associated with relational, more-than-human modes of ethics. What was necessary in the context of the trial was not creating a more nuanced articulation of the issue at stake but working to actively contest existing structures that undermined the possibility of this articulation. Indeed, after the trial, the first aim of the activists was to actively attempt to transform the legislative structures that had denied them aid, through challenging the UK legal system itself in order to prevent wealthy actors from repressing marginalized standpoints in the future (a challenge they won, with the European Court of Human Rights ruling that the activists' rights had been violated when they were denied legal aid).[78] Similarly, the animal welfare portion of the trial resulted in a favorable outcome for the activists not just because they were able to trace a complex set of agricultural, industrial, and legislative relations but because the activists drew attention to the ethical potentials that were excluded by these existing arrangements. These examples, in other words, were attempts to draw attention to the realities foreclosed by existing sociotechnical relations, as a means of arguing that these relations needed to be contested.

What the McLibel trial thus foregrounds is a dilemma wherein both representational modes of rights and more nuanced modes of articulation—which try to capture the irreducible complexity of a given situation—offer problems. In the theoretical context at stake here the problems with representation are well established, as with Haraway's powerful critique in "The Promises of

Monsters." In contrast, issues surrounding relational modes of ethics tend to be framed as matters of implementation. In his overview of relational theories, for instance, animal geographer Henry Buller suggests that work to support the "political expression and mobilization of this emergent relational ontology" is still ongoing.[79] As touched on in the introductory chapter, Alexis Shotwell, similarly, argues that it is often unclear how to decide between "relational economies" or how to replace relations deemed problematic, setting out a more concrete series of suggestions about how to negotiate these difficulties.[80] What such arguments point toward is a need to somehow work through the tensions associated with relationality in order to offer a firmer sense of the modes of politics it offers. While I am sympathetic to these arguments, my point is slightly different: perhaps what lies at the heart of tensions between these strands of theory and practice is something more fundamentally problematic about the types of ethics that emerge from entanglement and being-in-relation.

In beginning *When Species Meet* with a reference to the alter-globalization movement, Haraway suggests there is promise in activist "approaches to neoliberal models of world building" that "are not about antiglobalization but about nurturing a more just and peaceful other-globalization."[81] This opening sets the stage for her subsequent exploration of coconstitutive relations between species that work to unsettle exploitative or violent practices that are couched in a liberal-humanist ordering of relations.

What Haraway's characterization of the global justice movement fails to capture, however, are aspects of anticapitalist politics that fail to fit with narratives of relationality. Global justice groups often anchored their alternative models of globalization in decisively *anti*capitalist values. Tactics engaged in by activists during the McLibel trial had productive affinities with theoretical work, particularly their attempts to highlight points of friction as a means of denaturalizing the infrastructural arrangements fostered by McDonald's. What underpinned this politics, though, was a commitment—to revisit the opening quote from McSpotlight—to contesting what they saw as the "fundamental problem" of capitalism. Although the activists were pushing for alternative forms of social organization, they were also calling for the *exclusion* of others; indeed, this exclusion played a necessary and productive role.

In McLibel, for instance, what constantly guided activists' attempts to articulate the processes of enrollment and emergence of norms, in the context of the vast infrastructures of McDonald's, was the sense that an alternative way of doing things was possible. While the tactics of highlighting infrastructural frictions might have enabled them to realize these aims, these approaches were ultimately informed by anticapitalist commitments. As with the animal welfare issues or indeed the broader dynamics of the trial discussed above, drawing attention to what is excluded from existing ways of doing things can be a valuable component of activist practice. However, it is important to underline that these attempts to contest existing exclusions are coupled with alternative visions of the world that necessarily enact exclusions of their own.

As with Franklin Ginn's slug narratives or Jo Freeman's concern about informal hierarchies, therefore, what the McLibel trial illustrates is that often the articulation of alternative ways of doing things is entwined with the contestation of existing relations that inhibit the expression of these alternatives.[82] To word things differently: in order to explore how things could be otherwise (to again reiterate Star's words), it is sometimes necessary to push for these alternatives at the expense of the relations that currently undermine them. It is for this reason that an ethical focus on exclusion—or more specifically on the particular relations or ways of being that are foreclosed as others are materialized—deserves further conceptual attention. In the following chapters I take a lead from the issues raised here in order to flesh out the value of exclusion in more depth, before exploring more critical questions about how political responsibility can be taken for the exclusions that are bound up with any form of intervention.

2 | Uneven Burdens of Risk

Particular possibilities for (intra-)acting exist at every moment, and these changing possibilities entail an ethical obligation to intra-act responsibly in the world's becoming, to contest and rework what matters and what is excluded from mattering.
—Karen Barad, *Meeting the Universe Halfway*

How can we fight for a better world if we don't share our ideas and activities with those around us who live outside of our activist circles? What chance do we have if the vast majority of the "general public" either don't hear about our activities at all, or only from perspectives other than our own? . . . [W]hat were we actually trying to achieve in organising actions. . . . And how did this relate to questions of mediation and representation within a movement that holds dear its political diversity and its critiques of leadership and representation?
—CounterSpin Collective, "Media, Movement(s) and Public Image(s)"

The above quotations speak to a very particular point: in order to create situated knowledge, it is vital to pay attention to the tools that are entangled with this production of knowledge. Attention to the materialities of communication is precisely what was called for in the previous chapter, with its focus on the mundane constraints that inhibited what activists were able to accomplish. This chapter moves on from the strictures identified in the McLibel trial, instead turning to more concerted attempts by activists to *overcome* these constraints through developing tools better suited to articulating polyvocal anticapitalist protest.[1] Here I focus on alternative media networks that were developed by activists both as experimental attempts to create more open and participatory forms of communication and as a means of combating critical voices' exclusion from mainstream media environments. Although media experiments are provocative in and of themselves, they also offer insight into broader theoretical issues surrounding the politics of openness and experimentation, specifically how a conceptual emphasis on these qualities could inadvertently consecrate inequalities while appearing to do the opposite.

The wider importance of representation and mediation is underlined by the activist media collective CounterSpin in the above extract; what emerges as important is not just *what* is being articulated but *how* these articulations are crafted.[2] In particular, the collective emphasizes the relationship between activist media and prefigurative politics. It is vital, from a prefigurative perspective, that the communicative tools that support protest are not treated as neutral intermediaries or a means to an end. It must always be asked whether particular media configurations inadvertently foster relations (such as hierarchy and insularity) that undercut the world that activists aim to bring into being. For this reason, prominent anticapitalist groups from the early 1990s to the present day have experimented with digital media in order to prefigure desired forms of social organization though the very media technologies that are used to communicate about these desires.

These media initiatives evolved and expanded throughout the first decade of the twenty-first century and while they provided an important space for dissenting knowledge, problems also began to emerge. Although networks such as Indymedia played an important role within contemporary histories of alternative media, they also created new norms and, by extension, new hierarchies that were grounded in assumptions about what open, participatory forms of politics should "look like" in practice.[3] What these activist media initiatives illustrate, therefore, is the need to remain attuned to the way that dangerous new norms can arise not despite but because of experimental attempts to displace repressive structures with approaches that seem more open and inclusive.

The media experiments I focus on to develop these arguments are especially helpful for fleshing out the importance of centralizing exclusion. As illustrated by the opening quotation, for Karen Barad any apparatus that generates knowledge about the world is necessarily entangled with its object of study, and the nature of this entanglement has profound ethical, epistemological, and ontological consequences. Certain sociotechnical arrangements, for instance, can draw attention to particular ethical concerns, contest normative forms of knowledge, and make certain realities possible, but in doing so they foreclose other possibilities.

Synthesizing the issues raised by the above extracts, my concern in this chapter is with the sort of realities that are foreclosed—perhaps paradoxically—through risky and experimental attempts to create more open, participatory media infrastructures. As hinted at through this shared vocabulary of risk, experiment, and openness, these media initiatives seem to embody theoretical calls for cosmopolitical approaches to knowledge production.[4] The problems faced by these initiatives, by extension, help to complicate neat mappings of cosmopolitics onto praxis. To elucidate these issues, I focus on the rise and decline of two instances of activist media use that are related to the campaigning work outlined previously: the McSpotlight website and the Indymedia news network. The bulk of this chapter draws on observations from my own research (both participatory and documentary), synthesized with analyses from a range of social movement media scholars, to examine the frictions that have characterized these contemporary histories of activist media use. As I illustrate throughout the chapter, these difficulties are not inconsequential but pose important ethical questions about the uneven burden of risk that can arise from open and experimental ethico-political approaches, where the consequences of particular experiments are more dangerous for some actors than others.

The Importance of Mediation: Situatedness and Scale

Despite the problems that ultimately plagued alternative media experiments such as Indymedia, they were originally seen as important in addressing the long-standing problem of how to develop protest networks on a transnational scale. Scale, or more specifically the act of scaling up particular ways of doing things (be they political or industrial), is a dangerous proposition. For Anna Tsing, any attempt to expand something to a transnational scale is problematic as such projects must by nature "be oblivious to the indeterminacies of encounter."[5] Anything, in other words, that is locally specific, nonscalable, and resistant is seen as a hindrance to projects of scalability and something that must be removed. Tsing elucidates this point in relation to forestry, arguing that the globalization of this industry means that unwanted actors are often reshaped to make for easier scalability: "During the heyday of joint public-private industrial forestry in the 1960s and 1970s this meant monocrop even-aged timber stands. . . . Unwanted tree species, and indeed all other species, were sprayed with poison. Alienated work crews planted 'superior' trees."[6] Tsing suggests (echoing Susan Leigh Star's wariness of the norms and standards that come with large-scale sociotechnical networks, as discussed in the previous chapter) that businesses often operate on the same model, imposing one-size-fits-all models of doing things with the aim of maximizing profit.[7]

As detailed in the previous chapter, the diverse consequences of scaling up were precisely what were articulated by activists during the McLibel trial, where industrial processes of food production, the deskilling of workers (which came with the standardization of kitchen equipment), and the alleged erosion of local food cultures all came under attack. Indeed, these scaling-up processes are what have led to McDonald's being cited as an especially problematic instance of corporate expansion, as reflected by more recent narratives of McDonaldization.[8] Activists, then, were faced with a dilemma: in order to respond on the transnational scale required, it was necessary to articulate the relationships among issues ranging from rain forest destruction to workers' rights. Yet working at this scale is often problematic because it relies on reductive one-size-fits-all models of dissent, which necessarily exclude perspectives and practices that do not fit with this model.

Scale has also been seen as a problem for knowledge production; the need to resist scaling up (and the abstractions that come with it), for instance, informs Donna Haraway's insistence on situated ethics and speaks to the longer tradition of situated knowledges that has been central to feminist science studies. Situatedness has, in other words, been seen as a means of avoiding the smooth scaling up of *ethical* frameworks that (in Tsing's terms) are "oblivious to the indeterminacies of encounter," indeterminacies that might unsettle anthropocentric or ethnocentric ethical norms. The difficulty faced by non-anthropocentric theoretical work, therefore, is finding a way to recognize and contest the consequences of anthropogenic activities that have global ecological consequences, while resisting equally totalizing ethical solutions.

The "ethics of storytelling" that has been called for within the environmental humanities has been positioned as a fruitful means of articulating situated entanglements of the local and the global, as embodied by Thom van Dooren's careful storying of the lives of birds or Tsing's close attention to the capacity of mushrooms to thrive in forests decimated by industrial processes.[9] Though particular in their focus, these stories evoke how the lives of specific species, in specific contexts, are bound up with global processes, such as the Pacific currents that break down discarded bottles into microplastics, which slowly poison albatrosses; or the flows of commerce that bind small, migrant mushroom-picking communities in the Pacific Northwest to fine-dining restaurants in Japan. In making the lives of particular species and communities visible, these stories pose urgent questions about the implications of specific knottings-together of different ecological scales. Storytelling, therefore, enacts the situated approach called for in the environmental humanities, by providing a rich, detailed account of multispecies entanglement, while bringing sharply into focus what is at stake when threads that seem to be irrevocably knotted begin to fray.

In *Staying with the Trouble*, Haraway links storytelling to protest in a more overt sense, focusing on the role of stories in actions undertaken by Indigenous activists in resisting mining activities and related water depletion in Black Mesa. Integral to these protests, Haraway argues, is using careful storytelling practices at a local level, to connect situated cultural meanings to specific scientific knowledges. The value of drawing together these different knowledges, she argues, is in order to build careful alliances between communities, which can then act as a foundation for a broader transnational solidarity network oriented around questions of water justice.[10] Two aspects of her account of protests are thus especially important: it is not just that stories need to be rich, detailed, and situated but also that attention needs to be paid to the *process* of weaving together stories from different communities and forms of knowledge, to ensure that this does not reinscribe hierarchies but instead acts as the foundation for broader alliances.

What is insisted on in these theoretical arguments is the importance of adopting a cosmopolitical approach to producing knowledge, of the sort advocated by Isabelle Stengers. For Stengers, it is crucial to respond to urgent ecological issues without perpetuating hierarchies of particular forms of expertise. Her cosmopolitical proposal marks an attempt to overcome hierarchies in knowledge production by creating space for dissenting knowledge to be "heard 'collectively,' in the assemblage created around a political issue."[11] The task, following Stengers, is thus to experiment with techniques for ensuring that those affected by a particular issue are not only heard but responded to. This form of openness and—as Haraway phrases it in her own application of cosmopolitical theory—"response-ability" toward voices and perspectives that are often excluded is designed to open up the risk of having to entirely reevaluate existing ways of doing things. Although such approaches are therefore risky, for Stengers they are nonetheless critical in building common worlds. The affinities between Stengers's arguments and anticapitalist activism mean that it is perhaps unsurprising that activist initiatives—and alternative media in particular—have often been framed as modes of cosmopolitics (and, as outlined in the next chapter, that these groups have in turn inspired Stengers). Yet a closer look at media activism also foregrounds the danger of experimenting with more open and responsive forms of politics; even openness is constituted by particular exclusions.

The riskiness of experimenting with practices and techniques for collective knowledge production, in part, is that such approaches can unsettle normative ways of knowing. For Stengers (and others who have engaged with her work), this riskiness, then, is something to embrace. However, experimental practices of knowledge production can also carry other, far less desirable forms of risk. As social movement scholars Paul Chatterton and Jenny Pickerill argue, anticapitalist activists regularly engage in risky and experimental forms of political organization in order to articulate links between the local and the global. The problem is that the risks of these approaches are not always borne evenly; indeed, one of the main conclusions Chatterton and Pickerill drew from their multisite ethnographic research into autonomous activism was that, despite the success of local initiatives, "transnational solidarities remain[ed] an unfulfilled ideal for many activists."[12] In practice, the construction of links between the local and the global was fraught with difficulties because "telling convincing narratives linking specific places to their wider context relies on experienced and skilful narrators."[13] The danger of narrating links between different struggles, therefore, is that this runs the risk of privileging small groups of transnational activists who are perceived to already hold the necessary expertise for accomplishing this task, while silencing those who are perceived to lack this expertise.

Although in the 1990s and early 2000s digital media technologies were seen as promising tools for experimenting with more open and responsive forms of communication and collaboration, these experiments with openness were predicated on some discomfiting exclusions that mirrored the sort of issues identified by Chatterton and Pickerill. The push for openness, and for new ways of ensuring disparate voices were heard, at times created a vacuum that led to the reemergence of informal hierarchies. In the case of the initiatives discussed here, these hierarchies were especially dangerous in reinscribing explicitly raced, gendered, and classed social relations. Both the potentials of engaging in experimental modes of politics and the exclusions they can give rise to come to the fore in the cases of McSpotlight and Indymedia.

Early research on activist media networks stressed their "deterritorializing" capacity to traverse national boundaries and legislative frameworks.[14] This point was, to an extent, true of McSpotlight, whose use of mirror sites outside the United Kingdom offered a means of overcoming UK libel laws, a point noted by prominent texts such as Naomi Klein's *No Logo*.[15] While especially high-profile, Klein's analysis of McSpotlight is just one instance of a broad range of research that emphasized the transnational significance of the site. Indeed, a focus on anti-McDonald's campaigning reveals over twenty years of experimentation with digital media, in addition to the campaign's sustained engagement with different printed materials from the 1980s onward, which makes it a privileged movement for examining the implications of this sort of experimental approach.

As discussed previously, McLibel (and McSpotlight) foreshadowed broader protest networks associated with the global justice movement, which gained substantial media and academic attention due to high-profile coordinated protests, such as 1999's Battle of Seattle, when activists converged in Seattle to protest the World Trade Organization. Drawing together groups affected by the excesses of globalization, from workers' groups based in Seattle itself to solidarity groups associated with the Zapatista National Army of Liberation (themselves a key source of inspiration for anticapitalist protest throughout the 1990s and early 2000s), the protest gathered a huge amount of academic, popular cultural, and mass-media attention.[16] From the perspective of activists and social movement scholars focused on activism, the protest was seen as heralding the birth of a movement that did not speak for those affected by globalization in different global contexts but attempted to work with them. "Working together," however, did not equate to different groups adopting abstract overarching aims that ignored the asymmetrical conditions particular communities might be working within.[17]

In his contribution to *Shut Them Down!*—a collection of essays about protests against the summit of a different transnational institution, the G8—social movement scholar Rodrigo Nunes, for instance, argues that networked technologies were initially heralded as holding the capacity to make nonhierarchical political organizing "materially possible on a large scale."[18] Nodding to autonomist Marxist perspectives, he suggests that—for some of the more prominent attempts to craft global anticapitalist networks—the ambivalent qualities of global, informational capitalism offered important political potentials. The potentials of digitally enabled networks were underlined by the CounterSpin Collective's aforementioned contribution to the same volume of essays; they prefaced their reflexive discussion of the difficulties of articulation by stating, "One of the key achievements of the last ten years has been the establishment and development of our own forms of media. The global Indymedia network, SchNEWS, wiki-based websites, independent films and local papers are all examples of attempts to create multiple spaces for debate, and for sharing information about our activities and politics with each other and with the wider world. *More than mere sources of 'news,' these media have provided vital inspiration, helping to build a real sense of solidarity amongst our many diverse (and diversely located) struggles.*"[19] Digital media technologies were, from this perspective, seen to support some of the integral practices of nonhierarchical social movements in facilitating processes such as consensus decision-making on a large scale, while preserving the more decentralized forms of organization associated with small, structureless affinity groups.

McSpotlight was an early example of the potentials attributed to the internet in particular; launched in 1996, the website was repeatedly cited as an example of digital media's capacity to support radical praxis, in this instance by offering communicative tools for activists in David-and-Goliath struggles between campaigners and multinational corporations.[20] What was important about McSpotlight, however, was not just its capacity to afford visibility to anti-McDonald's campaigning but its ability to achieve this in a way that facilitated dialogical engagement at all stages in the composition of the issue. The website was launched, in other words, to overcome the sorts of constraints identified in the previous chapter, during the course of the McLibel trial, which made it difficult for activists to articulate the complex relations that existed among disparate issues.

On the most basic level, the site offered a space for the multilayered articulation of McDonald's, which was crafted not just by London Greenpeace activists but by workers and activists based in other global contexts during the course of the McLibel trial. In addition to archiving evidence from the trial, the site also continuously updated and developed its narrative, through including stories from actors who were precluded from attending the trial itself due to the expense, and through embedding more dialogical spaces for those most affected by McDonald's to discuss and even contest the narrative constructed about the corporation (such as web forums dedicated to McDonald's staff). These features might, in a context of social media and corresponding suspicion of the political potentials of digital platforms, seem dated or ineffectual, but what the site's development brought to the fore was the need to consider not just the composition of issues in themselves but the tools through which articulations are made, in terms of their capacities to create spaces for different perspectives to enrich or indeed unsettle activist assumptions.[21]

This sort of appraisal of digital media might seem overly optimistic in the contemporary political climate, but it is important not to mischaracterize early activist (or indeed academic) focus on the internet as entirely uncritical. More contemporary, critical narratives about digital media tend to construct a particular story about the scholarship of digital culture that presents early analyses as overly utopian (and themselves as a welcome critical redress to this optimism). However, this sort of neat mapping can flatten out important debates.[22] Within social movement studies in particular, analyses of activist media displayed considerable nuance in terms of how the potentials of digital media were conceptualized. Pickerill's detailed study of McSpotlight, for instance, elucidates the value of digital alternative media for environmental campaigns in the late 1990s, while contesting theoretical claims that suggest activists consistently put their faith in "techno-fixes."[23] Instead, the activists understood their "computer use . . . [as] a compromise in an imperfect world."[24] For instance, in the case of McSpotlight, there was keen awareness that a reliance on digital media could perpetuate, or even create, forms of inequality: "Many accepted that the accessibility of CMC [computer-mediated communication] was limited and that there were exclusions: 'you're excluding the poor, the off-lined, the people in countries that don't have great internet access' (Gideon, McSpotlight)."[25]

Similar reflections emerged in relation to the administration and development of the site, with activists concerned that a shift to the digital would create informal hierarchies by privileging those activists with technological skills. Due to being a pre–Web 2.0 website, McSpotlight still relied on a production model that required particular activists to assume a gatekeeper role: in addition to the architecture of the site being developed by those with expertise, all information had to be directly uploaded by a group of individuals. There was further reflection about problems of male dominance in the information technology industry (which had an impact on the demographics of volunteers), concern about the working conditions of those who produced information technology system components, and acute reflection about the environmental consequences of communications technologies.[26]

The problems that activists attributed to computer use, moreover, were not naively accepted as "the way things are"—a necessary sacrifice for scaling up—and were instead acted on.[27] McSpotlight was a project that involved constant work and experimentation with different forms of organization, in order to negotiate emergent problems. This ongoing work can be understood as what Hilde Stephansen and Emiliano Treré term "capacity-building" practices.[28] *Capacity* has a dual meaning in the context of activist media networks, referring both to practices employed in activism in order to enhance people's capacities to participate actively in media making, and to the way that these practices go on to expand the capacity of the networks themselves. Both of these forms of capacity building were evident in McSpotlight. In line with the importance of skill sharing to anticapitalist politics more broadly (a feature of practice discussed in more depth in the next chapter), activists experimented with a diverse range of capacity-building practices to overcome everyday tensions that arose in the maintenance of McSpotlight.[29] For instance, workshops to teach people basic HTML and broaden participation were held, and materials were published in multiple languages to overcome linguistic hegemony. Though McSpotlight activists (and a number of the members of other groups interviewed by Pickerill) recognized the environmental and labor problems associated with the computer industry, moreover, they strove to avoid the logic of built-in obsolescence, making do with older hardware rather than upgrading regularly.

Awareness of the hierarchies that could emerge due to the uneven (often classed and gendered) distribution of technical skills also meant that activists were wary of privileging one particular media platform over others. Pickerill makes it clear that the site needs to be situated in relation to other media engaged in by activists: "just one aspect of a varied strategy co-ordinated by the McLibel Support Campaign," among other approaches including "mass leafleting, media focus, pickets outside McDonald's stores and international days of action, with links to residents' opposition groups and disgruntled McDonald's workers."[30] Even amid early hyperbole about the politics of network technologies, therefore, digital media were not seen as a solution for unproblematically expanding anti-McDonald's activism to meet the international scope of the campaign, but as something that needed to be used with care and underpinned by sustained capacity building in order to avoid hierarchical advocacy relations.

What McSpotlight helps to make visible, therefore, is that openness requires constant *work* and constant tinkering with (in this instance) the media arrangements at stake. My own research, conducted ten years after Pickerill's, reveals a similar experimental ethos associated with the work of attaining openness. Even though activists continued to engage with many of the media described by Pickerill, the ecology of practices associated with these media had changed in ways that subtly reshaped their affordances. What was revealed through examining these changes were some subtle practices of negotiation with the imperfect affordances of particular media, something evident in relation to paper pamphlets as well as digital media.

As discussed previously, sometimes pamphlets can inadvertently slip into representational forms of advocacy that instill hierarchical relations between activists and those they seek to represent. In her work on the media practices of anticapitalist activists, Veronica Barassi argues that paper pamphlets often emerge in grassroots settings, with their content decided on dialogically.[31] What becomes apparent in the case of a long-standing campaign such as the anti-McDonald's protest is that pamphlets' dialogical origins can become obscured over time due to problems that, on the surface, seem fairly innocuous. Printed pamphlets often languish in radical libraries or in boxes of protest literature and are still distributed years after their original compilation. In relation to McLibel, for instance, pamphlets printed in the mid-1980s continue to be circulated today.[32]

The *What's Wrong with McDonald's?* pamphlet was produced by Veggies in the early 1980s and continues to be one of the most prominent pamphlets associated with anti-McDonald's campaigning. Though the pamphlet coexisted with the fact sheet at the center of the McLibel trial, it went on to supersede the original leaflet because it offered a more succinct summary of the issues at stake. *What's Wrong with McDonald's?* rapidly became the main pamphlet for distribution on the McSpotlight website, with copies available in seven languages, and two million pamphlets downloaded in the five years following the site's launch.[33] The leaflet also underpinned the activists' "adopt-a-store campaign," with physical and virtual copies freely distributed as resources for local activist groups to picket their own branches of McDonald's.

Both pamphlets were originally designed to "gather up" stories of McDonald's that were produced by different activist groups, in order to craft connections across different contexts, issues, and experiences. Once protest literature assumes a fixed form, however, it can become not only normative but detached from the experiences of those it seeks to represent, a problem illustrated by the succinct format of *What's Wrong with McDonald's?* in particular. I experienced some of these tensions firsthand at a local picket for the International Day of Action against McDonald's in 2007, when a pamphleteer was approached by an ex-McDonald's employee who voiced disquiet about the pamphlet's description of workers being "forced to smile." Of particular concern was the inference that workers were not smiling out of their own will but due to corporate imperatives, which the worker felt was a misrepresentation of their own actions. On another occasion, in 2010, members of the public cast doubt on claims made in the pamphlet about animal welfare, due to the recent launch of a new marketing strategy by McDonald's UK that promoted its use of British produce. These brief encounters are typical of the low-level contestation of pamphlet content that occurred regularly at protests. These incidents, more profoundly, speak to the way that a political semiotics of articulation can easily slip into a more decontextualized representation of broad problems, due to everyday practices associated with the distribution of protest literature.

In light of the initial excitement over McSpotlight in the late 1990s, it would be tempting to assume that these issues can be resolved through using more participatory online media. This was not the case in practice, however, as the problems associated with pamphlets were not overcome through simply displacing them with online media but through staging the *What's Wrong with McDonald's?* pamphlets in subtly different ways. A key development in the groups I was involved with in Nottingham, for instance, was the decision to hold protests on the anniversary of the trial, thus historicizing the campaign in such a way that the pamphlet served primarily as a marker of defiance against corporate attempts to quash protest. In addition to being situated in relation to a specific protest history (in which the local activist community had played a key role), the pamphlets were distributed along with new protest literature—including a pamphlet entitled *McGreenwash*—that unpicked misleading claims made by McDonald's UK in their rebranding. Through these seemingly trivial acts of recontextualization, the pamphlet was repositioned within a specific lineage of local protest, and a specific national context, in order to nuance the broad historical claims it made about the corporation's activities.

McSpotlight itself was similarly recontextualized. From 2005 onward, when Indymedia was launched in Nottingham, the latter website had been steadily used to document local activism, and by the time of my own involvement, it was the main source of alternative media used locally, with anti-McDonald's protests both documented and publicized on the site (indeed, I regularly participated on the site during this period in documenting particular protest actions).[34] McSpotlight, however, was not entirely displaced by Indymedia but instead assumed a different function, acting as a source of historical information about McDonald's as well as an archive of protest history. Indymedia reports about local McDonald's protests consistently linked to McSpotlight in order to narrate local actions as part of a more enduring, transnational movement history.

Over time, therefore, the media described by Pickerill were not treated as straightforward awareness-raising platforms (in the manner that they had been originally) but were instead carefully recontextualized and rearticulated as part of a specific transnational protest history. This negotiation of the constraints and contradictions of particular media continues in contemporary campaigning work; as I have described in my previous research on digital food activism, in their more recent food campaigning activities (discussed in depth in the next chapter) Veggies juxtaposes a range of different media on its website. Via what Adam Fish describes as the mundane "remix aesthetic" of digital culture, various media are brought together on their food-activism web pages, including links to old Indymedia reports of protests, PDFs of pamphlets, and the feed of their Twitter account.[35] Through situating social media as part of a more sustained protest narrative, activists were able to ameliorate the problems of ephemerality and superficiality associated with platforms such as Twitter. By embedding social media within their web pages in such a way that this content was juxtaposed with "older" media, Veggies was able to subtly reshape social media's affordances from being the sort of fragmentary format derided by critical scholarship to serving as just one component of a much more complex and sustained media environment that combines different platforms to articulate and sustain collective action.

Overall, therefore, in an examination of media use in anti-McDonald's campaigning, what is foregrounded is how these everyday but nonetheless complex uses of pamphlets, websites, and alternative and social media have to be underpinned by substantial work in order to build capacity and contextualize protest actions. What these negotiations illustrate is that openness requires constant work, attention, and experimentation, which cannot be delegated to media technologies themselves. To put things in Stephansen and Treré's terms, it is instead the capacity-building work that surrounds these media that is critical. The conclusion that could be drawn from these anti-fast-food media environments, then, is that constant experimentation with building capacity is needed so that no one model congeals—to put things in Nunes's terms—as what openness "looks like." The role of experimentation within these media initiatives offers some productive connections with cosmopolitical thought; hence, it is perhaps unsurprising that as well as being interpreted *as* cosmopolitical, particular activist movements have been engaged with by Stengers herself. My positive appraisal of these tactics also points to certain dangers, as well as the need to resist uncritically valorizing any approach, even risky experimental ones. A different anticapitalist media experiment, Indymedia, for instance, both complicates any optimistic conclusions that can be drawn about experimental praxis and elucidates the particular dangers if the partners involved in these experiments are working on uneven ground.

Indymedia: Sociotechnical Norms and Informal Hierarchies

Indymedia assumed an even more privileged position than McSpotlight in academic narratives about digital media and activism, and is often described as the archetypal activist media "experiment" or radical-participatory "laboratory."[36] Indeed, during the early 2000s the network was one of the most heavily discussed examples of attempts to craft a large-scale anticapitalist protest infrastructure.[37] On June 18, 1999 (to coordinate with the twenty-fifth G8 Summit in Cologne), a series of anticapitalist protests, dubbed the Carnival against Capital, took place in cities including London, Port Harcourt, Barcelona, Melbourne, Tel Aviv, Prague, Toronto, Rome, and San Francisco (in addition to Cologne itself). In London, in order to contest depictions of violent protesters that overshadowed the political stakes of these protests, activists experimented with digital media to document protest events in ways that offered a clearer articulation of anticapitalist values.[38] Following these on-the-ground experiments, the first Indymedia center was launched later that year in Seattle to provide a platform for anticapitalist perspectives at the high-profile series of protests against the World Trade Organization.[39]

Seattle was the starting point for a larger radical-participatory media network that not only drew together local news sites (each of which published entirely user-generated stories) but was entirely run, developed, and organized by users. As with McSpotlight, nonhierarchical communications infrastructures were often difficult to realize in practice because certain groups or individuals ultimately had to facilitate site development; in this case, local Indymedia centers were supported by editorial collectives, which were often associated with the autonomous social movements that had given birth to Indymedia.[40] The purpose of the network was not to be activist-only (with all the potential this brought for compounding the marginalization of particular groups from the media) but to provide a platform from which communities who were ordinarily excluded from the mass media could speak. In Nico

Carpentier, Peter Dahlgren, and Francesca Pasquali's terms, by encouraging participation in media organization, Indymedia was "radically participatory," rather than just promoting the principles of nonhierarchy through their content.[41] Throughout the early 2000s Indymedia seemed to meet these participatory aims and was celebrated as a political success story; centers flourished in locations such as Chiapas and Palestine to support the work of grassroots community activists, for instance, and Australian media activists engaged in sustained outreach to involve Aboriginal communities in the running of Indymedia (to name just a few examples).[42]

Yet, as early as 2007, commentators were describing Indymedia as a "failure."[43] Assertions of failure were perhaps premature (and certainly presumptuous with regard to what constitutes failure), with the network continuing to thrive in certain regions until the early 2010s.[44] These caveats aside, my own research into Indymedia's development nonetheless revealed a sharp decline in the number of active centers post-2010, from 175 at the network's peak to 68 by 2014 (with the number of active centers since declining still further), and no active centers at all in Africa, or East or South Asia.[45]

When I explored the reasons for this decline, it became apparent that a diverse range of factors were involved, but what proved especially damaging was when certain conceptions of how to move beyond hierarchy—which had proven successful in particular contexts—congealed as models for what radical democracy should look like. Indeed, a number of thinkers have pointed out that the specific model of horizontality developed by the network (which valorized consensus decision-making and openness) gave rise to informal hierarchies, the sort of invisible inequalities identified by Jo Freeman, which persist in apparently structureless groups and reproduce wider social and cultural inequalities.[46] The emergence of hierarchy has been a persistent problem for groups that aspire to horizontality; as Freeman has argued in relation to the use of structureless groups in women's movements, sometimes a lack of structure masks the continued existence of informal hierarchies:

> The idea [of structurelessness] becomes a smokescreen for the strong or the lucky to establish unquestioned hegemony over others. This hegemony can easily be established because the idea of "structurelessness" does not prevent the formation of informal structures, but only formal ones. . . .
>
> Because elites are informal does not mean they are invisible. At any small group meeting anyone with a sharp eye and an acute ear can tell who is influencing whom. The members of a friendship group will relate more

> to each other than to other people. They listen more attentively and interrupt less. They repeat each other's points and give in amiably. The "outs" they tend to ignore or grapple with. The "outs"' approval is not necessary for making a decision; however it is necessary for the "outs" to stay on good terms with the "ins." Of course, the lines are not as sharp as I have drawn them. They are nuances of interaction, not pre-written scripts. But they are discernible and they do have their effect. Once one knows with whom it is important to check before a decision is made, and whose approval is the stamp of acceptance, one knows who is running things.[47]

These sorts of informal hierarchies can become intensified in the context of particular uses of technology, which are assumed to secure structurelessness or—in the case of Indymedia—openness.

Openness was manifested in a range of different ways within Indymedia; not only were the editorial meetings open to anyone who wanted to be involved, but activists engaged in concerted attempts to reach beyond the immediate activist context, to ensure the centers were fundamentally shaped by marginalized voices: indeed, having clear outreach plans was a prerequisite for a new center joining the network.[48] Despite the good intentions that lay behind them, Todd Wolfson argues that these manifestations of openness became a problem, due to being supported by practices that were directly transported from the revolutionary Zapatista movement.[49]

Based in Chiapas, Mexico, the Zapatistas grew to prominence in 1994 and declared war on the Mexican government, in part in response to the effects of the North American Free Trade Agreement (NAFTA) on Indigenous land and livelihoods.[50] They were a particularly high-profile movement that articulated dissent in explicit response to neoliberal policy and pioneered the use of the internet to draw international attention to their cause. For this reason, it is difficult to overstate the movement's influence on, for instance, anticapitalist protest throughout the 1990s and early 2000s, activist engagements with digital media more specifically, and—in academic contexts—social movement studies, media studies, and cultural theory more broadly.[51] Indeed, Haraway more recently cites the Zapatistas as an ongoing source of inspiration in *Staying with the Trouble*.[52]

Due to their extensive influence, there is not space here to go into depth about the dynamics of Zapatista philosophy, *Zapatismo*, or the history of the movement, at least not in a way that does their work justice, but it is nonetheless important to underline that the work of the Zapatistas has proven vitally important for a diverse range of social movements, as well as a diverse range of academic work. What Wolfson points out, however, is the dangers that can arise when the situatedness that characterizes Zapatismo is lost, a problem brought to the fore by Indymedia.[53] Wolfson argues that the particular models of openness and consensus that informed the organization of Indymedia were directly transported from Zapatista praxis.[54] The problem with taking inspiration from Zapatismo, for Wolfson, however, was the way these principles were scaled up, which led to problems. He argues that there were "considerable differences between the nature of the geographically bounded [Zapatista] struggle in Mexico and the development of a transnational alternative communications network for social movements. The Zapatista model of revolution was flexible and driven by the historical and material contexts they were facing in Mexico. Indymedia activists took the products or outcomes of this model as rigid laws, and, therefore, missed the spirit of the Zapatista praxis as it was a dialectical, adaptable approach to struggle."[55] Principles of openness, Wolfson goes on to argue, became a particular problem when "attempting to make clear decisions across space, time, and language barriers."[56] While these criticisms perhaps downplay the reflexive approach taken in particular contexts by Indymedia activists, who both recognized and sought to overcome these problems (with Pickerill's "'Autonomy Online'" again offering a sense of some of these negotiations), some of the issues Wolfson identifies are nonetheless borne out in the difficulties faced by particular collectives.

A number of scholar-activists have argued, for instance, that the rejection of hierarchical values and the promotion of open, horizontal ways of organizing were seen as guaranteeing a nonhierarchical reality; that is, particular material arrangements were conflated with certain ethical relations. Fabien Frenzel and colleagues foregrounded significant problems in African contexts, where assumptions about what openness "looked like" in technical terms became especially exclusionary. The Indymedia Mali collective, for instance, was criticized for using the facilities of nongovernmental organizations and government buildings (even though this was a result of resourcing issues). The rigid way openness was understood and adhered to in this context thus reproduced precisely the colonizing relations the network was designed to overcome, wherein groups in more privileged contexts leveled judgment or imposed ways of doing things on others.[57] Similar tensions were found in relation to other Indymedia Centers; Frenzel and colleagues go on to describe, for instance, how it was suggested that activists in Nairobi "relocate to the slums" for meetings in order to meet the network's demands for open participation.[58]

Though well intended, such requests were seen as "glorifying poverty" and downplaying the specific struggles activists would face in accomplishing this task, in comparison with the outreach work faced by activists in Europe and North America.[59]

My own work, similarly, has explored how attempts to launch Indymedia Cairo and El Paso were stultified in 2011, in part due to bureaucratic approval mechanisms that were not sufficiently sensitive to the particular geopolitical constraints faced by the activists concerned.[60] In line with broader attempts by autonomous movements to engage in nonhierarchical decision-making, any decisions about the direction of Indymedia had to be made by consensus; at the local level this entailed consensus being reached by editorial collectives, with broader decisions about the network only realized via painstaking discussions among participants at the global level after consensus had been reached through intensive email-list debates.[61] In the case of the proposed centers in Egypt and El Paso, tensions existed between the need to have the centers prove their adherence to Indymedia's overarching principles of unity and the urgent need to launch the centers as soon as possible in order to circumvent media censorship.

Frictions were not just geopolitical, however, and Scott Uzelman's reflections on the decline of Vancouver Indymedia clarify the danger of valorizing particular models of openness still further. He describes, for instance, how oppressive social relations were perpetuated at a local level when allegations of sexism were dismissed because excluding the individuals at stake was against the collective's open values.[62] Uzelman goes on to argue that these problems could be attributed to two issues: first, activists "confused open access regimes with democratically regulated resources (i.e. commons)"; and, second, "some within the collective reified the technologies we used."[63] Again, these experiences in Canada illustrate the danger of reifying network technologies instead of the capacity-building practices necessary to sustain them.

These are a broad range of problems that manifested differently in different contexts, but what they illustrate are the specific ways that material arrangements can create (or foreclose) possibilities for collective knowledge production, in ways that have sharp ethical consequences. Though digital media seemed to offer tools for scaling up nonhierarchical values that had originally been employed in smaller group contexts, their capacity to instead create and consecrate preexisting social hierarchies can be masked when these tools are perceived as straightforwardly translating those values. When grappling with these problems, what is needed is to go beyond—to put things in Bruno Latour's terms—understanding media technologies as mediators rather than intermediaries (though this is important). In line with Star, what should be acknowledged in contexts such as this, I suggest, is the uneven distribution of the *work* involved in materializing experiments with openness, norms that (in the case of Indymedia) were virtually inaccessible to groups working in certain geographical and/or political contexts.[64] The hierarchies emerging in activist media environments, in such instances, can be seen as especially pernicious due to being implicitly gendered, classed, and racialized.

Problems with technological determinism aside, what also needs to be addressed are the inequalities fostered by an experimental ethos itself. In order to fully conceptualize the stakes of these complex problems, it is useful to turn back to the arguments introduced at the beginning of the chapter and revisit Barad's point about the importance of contesting and reworking what matters and what is *excluded* from mattering, in order to bring this point to bear on the technological arrangements that support protest networks. From Barad's perspective, any given assemblage brings certain realities into being and excludes others. In light of the geopolitical and gendered hierarchies that were inadvertently reproduced by Indymedia, I suggest, Barad's insights provoke important questions about what types of exclusions are bound up with the very experiments with openness that are often intended to overcome these problems.

Who Bears the Risk of Media Experiments?

As with its broader uses in figuring more-than-human worlds, entanglement has become an evocative concept within contemporary media theory, one that conveys something profound about the role of digital media technologies within the complex tapestry of relationships that constitutes the social. Pip Shea, Tanya Notley, and Jean Burgess, for instance, use entanglements and frictions as coordinates for identifying the complex role of digital technologies within protest movements (as well as the tensions surrounding this role) in their introduction to a special edition of *Fibreculture* that draws together analyses of the role of social media in contexts including protest camps and disability activism.[65]

The significance of entanglement as a framework for understanding activist engagements with media is still more explicit in Anna Feigenbaum's account of the media ecologies of protest camps, when she describes how the varied media employed in these camps play a messy, coconstitutive role that informs some of the fundamental characteristics of these camps. Media are crucial in articulating the aims of the camp to those outside of the immediate activist community, but they are just as important in supporting the identity formation of protest campers. At the same time, the affordances of media emerge only by and through their relationship with other protest camp actors—from activists to camping equipment—and in relation to the practices that become associated with them. In making these arguments Feigenbaum thus speaks to a contemporary lineage of work in media that has engaged with an understanding of entanglement that is informed by new materialist approaches.

What entanglement conveys in political contexts, such as protest camps, is that the relationships among specific people, protest cultures, and media technologies cannot be accounted for solely by noting the particular affordances of the actors involved, then reflecting on the ways that they coshape one another. In Barad's terms the affordances of any actor are not preexistent but emerge only through the actor's relationships with other entities. From her agential realist perspective, stable identities (including subject/object) emerge only through *intra-active* relations, a neologism she coins to contrast with *inter-action*. Everything from the traits of an atom to the labor relations of a factory, for Barad, can only emerge intra-actively through the sets of relations that constitute the assemblage at stake.[66]

The implications of Barad's arguments for activist media are elucidated by Feigenbaum's argument that the media used in protest camps are often objects that are not conventionally thought of as media. While email lists, pamphlets, and social media might all play a role in shaping praxis, so too do unexpected entities such as tents and tear gas canisters. These "other media" decisively serve a communicative function: tent canvases are often crucial in communicating the language of occupation to the public, and remnants of tear gas attacks are shared among activists "as people photograph and document branded labels of tear gas canisters, creating chains of global accountability."[67] The affordances of these (perhaps unlikely) media entities, however, are not determined in advance but emerge only through the specific protest ecology of the camp. Feigenbaum's arguments do not just apply to "other media," moreover, as even more conventional media technologies assume their properties from longer histories of practice and are shaped by the other media they are used in conjunction with.[68]

Yet while activist media's status *as* media might not be a given, because the properties of any medium (or indeed any entity) emerge only through its intra-actions within a particular assemblage, this is not to say that the properties of particular media technologies do not have *consequences*. Tents and tear gas, for instance, mediate protest in profound ways, as do the range of other technologies engaged with by activists in other contexts. In relation to activist uses of media, therefore, the significance of Barad's argument is twofold. First, her arguments provide a framework for conceptualizing how the affordances of media are the product of particular sets of relations at a particular sociohistorical juncture. Second, at the same time, Barad's approach emphasizes that—once these affordances have stabilized (in a similar manner to the processes of stabilization that Star described in the previous chapter)—they have material consequences that are difficult to reverse.

In the context of activist media use, Barad's specific understanding of entanglement helpfully encapsulates the irreducible nature of the relationships among media arrangements, practices of articulation, and modes of mobilization.[69] These arguments, moreover, also mark the similarly irreducible entanglements among the ontological, epistemological, and ethical dimensions of activist media ecologies. The stakes of this point can be elucidated in a general sense if related back to Indymedia and McSpotlight, where it was impossible to separate openness as a material, technological characteristic (as with open-source software and participatory hacker practices), openness with regard to the type of knowledges made possible (as with the collaborative, participatory dimension of radical alternative media), and openness as an ethical orientation (as with media experiments driven by concerns of nonhierarchy and inclusivity). Indeed, it was when the material affordances of these initiatives were treated as static, and decontextualized from the broader media ecology, that problems arose. The open, participatory affordances of particular media, in other words, cannot be reduced to the fixed technical attributes of particular technologies, as these affordances emerged only through the relationships among different practices, values, and materials: something that becomes explicit when reflecting on the significance of capacity-building practices to McSpotlight, or the problems faced by particular Indymedia centers.

At the same time, it is also important to reflect on the danger of reifying particular practices, even experimental ones; as Wolfson argues, and as elucidated by the issues I encountered in anti-McDonald's pamphleteering, the danger is in making assumptions not just about technologies but about a range of other practices that are assumed to secure openness. I argue that it is important to extend arguments from social movement studies about the danger of reifying particular technologies and practices still further, to ask questions about the danger of universalizing an experimental *ethos*. Again, Barad's work can offer a helpful route into these debates. Where her arguments differ from other conceptions of the performative nature of ontology (such as those offered by Annemarie Mol or Joanna Zylinska, for instance) is in her attention not just to the ways in which things are enacted but to the particular points within these assemblages when things stabilize as things are, which she describes as "agential cuts."[70] It is in these cuts that the ethics of Barad's work comes to the forefront, as this is the moment where particular realities and subject positions emerge and—crucially—others are foreclosed. This approach thus gives the material world contingent stability, while still opening space for things to be otherwise if different cuts were made. It is not enough to focus on the messy ways that media technologies, political situations, and protest actors dynamically coconstitute one another, then, but important to draw out the ethical implications of these entanglements by focusing on questions of exclusion.

Paying attention to exclusions is especially important when related to the concerns articulated in this and the previous chapter. What these chapters have foregrounded is that in order to understand the ethics of particular cuts, it is vital to recognize the work that is created when norms become stabilized. Echoing Star, the invisible work of negotiating sociotechnical infrastructures often falls unequally on those who do not fit with infrastructural norms. What this chapter has shown, however, is that this is a problem not just for those affected by large-scale sociotechnical infrastructures (such as McDonald's) but for activist initiatives. My central concern here has been how this uneven distribution of work is bound up with particular practices as well as technologies. Even though experimental approaches can be important in exploring how to redistribute this work in more egalitarian ways, it is essential to recognize that the act of experimenting with new techniques is in itself far riskier for some actors than others.

What I am not arguing, in drawing attention to these difficult issues, is that experimental practice should be dismissed. In the context of activist media, for instance, the legacies of Indymedia have been critically important for subsequent initiatives.[71] More broadly, experimental forms of knowledge production have made vital interventions in recognizing the messy terrain that all knowledge takes place within, while attempting to navigate a path through this terrain.[72] It is nonetheless still important to take responsibility for the way in which experimental approaches can, in certain contexts, reinforce preexisting inequalities, in order to find ways of overcoming this problem.

The media initiatives discussed throughout this chapter elucidate that intensely damaging relations cannot necessarily be combated by *more* openness or *further* experimentation, as in certain contexts these approaches can mask responsibilities rather than opening them up. Sometimes intersectional inequalities, fostered through particular relations, need to be more explicitly opposed. As stated in the previous chapter, Haraway begins *When Species Meet* by praising the work of alter-globalization activists and describes *alter-globalization* as a term "invented by European activists to stress that their approaches to militarized neoliberal modes of world building are not about antiglobalization but nurturing a more peaceful and just other-globalization."[73] As argued previously, what is missing from this characterization is that these activists also regularly used the allied term *anticapitalism* to mark the types of relations being opposed and excluded from their own practice.[74] To echo Freeman's arguments, it is a matter not just of creating space for others to speak and be heard, but of actively working to overcome and oppose affective relations that shore up existing oppressions.

While activism that seeks to move beyond hierarchy might aspire to prefigure alternative worlds, therefore, in order to achieve this it is often necessary to exclude or at least contest the relations that ordinarily inhibit these alternatives from coming into being. What is vital is ensuring that any form of exclusion happens in a way that is responsive and open to contestation. My argument, therefore, is not that risky, experimental approaches need to be avoided but that more work needs to be done to recognize and take responsibility for the uneven burdens of risk they can foster. The complications surrounding this task are the focus of the next chapter.

3 | Performing Responsibility

Inside rich histories of entangled becoming—without the aid of simplistic ideals like "wilderness, "the natural" or "ecosystemic balance"—it is ultimately impossible to reach simple, black-and-white prescriptions about how ecologies "should be." And so we are required to take a stand for some possible worlds and not others; we are required to begin to take responsibility for the ways in which we help to tie and retie our knotted multispecies worlds.
—Thom van Dooren, *Flight Ways*

While it addresses a single issue, animal liberation does pose fundamental questions about the relationship of humans to the world. This can be a starting point for a fundamental questioning of the way we live our lives; on the other hand animal rights ideology can become a limit which prevents a wider critique of society. We need to go beyond this ideology without abandoning what is subversive in what it represents.
—*Beasts of Burden*

This chapter is concerned with the difficulties inherent in, as Thom van Dooren puts it, taking a stand for certain worlds and not others. The *not*, I suggest, is particularly important. In environments where human lives are enmeshed with multispecies communities, it is often necessary, as van Dooren argues, to take some form of action in order to undo the damage wrought by anthropogenic activities. At the same time, it is vital to take account of the constitutive exclusions that underpin any form of ethical and political intervention, as any set of relations necessarily occurs at the expense of other possibilities. In order to craft relations of responsibility, in other words, it is necessary to be attentive not only to what is materialized through particular courses of action—the entangled relations that they bring into being—but to what is excluded.

This chapter confronts the difficulty of meaningfully taking responsibility for constitutive exclusions. While the need to take responsibility has become a recurring theme within relational theories, it often remains unclear how to actually realize these responsibilities. As I argue throughout this chapter, this lack of clarity has been intensified by rifts that have emerged between, on the one hand, academic and activist perspectives that have a specific commitment to action and, on the other, those that have stressed the need to refuse ethical solace in a world of irreducible entanglement and interdependency. For a number of thinkers, Donna Haraway chief among them, responsibility entails *response*-ability: a refusal to impose predetermined ethical solutions and an attempt to instead be continuously responsive to the needs and interests of relational partners.[1] For the anonymous author of *Beasts of Burden*, *responsibility* means something different: enacting a stand (in this instance, one committed to animal liberation) and then taking responsibility for its limitations by making them visible and open to contestation or transformation.[2] At stake, therefore, is whether adopting a decisive political stand necessarily delimits ongoing responsibility or can act as the foundation for it.

Within the chapter I turn to two sites in particular, protest camps and free-food giveaways, to take inspiration from a particular approach that has proven valuable in fostering responsibility on the part of activists toward the worlds they seek to bring into being: performative protest tactics. Performative protest is significant in illustrating how activists' ethico-political stands can be enacted in collaborative and highly visible (if messy) ways, which open space for these stands to be contested and renegotiated. These tactics are, in both contexts, supported by what can broadly be categorized as "responsibility practices," approaches that foster responsibility on the part of activists, not just toward the worlds that are temporarily enacted though their performances but toward those that are foreclosed.[3]

In taking inspiration from some of the creative ways that responsibility has been realized in activist practice, this chapter troubles existing theoretical accounts of what responsibility toward a particular stance might look like. In doing so, I aim to carve out space to recognize the interventional work that can be accomplished by more explicitly oppositional and critical perspectives.

The Politics of Openness: Ontological, Epistemological, and Ethical Surprise

In order to draw out the significance of responsibility practices, it is necessary to revisit theoretical debates about openness, experimentation, and surprise by situating them in relation to particular conceptions of ecology. The lens of ecological thought, in the sense offered by Félix Guattari, is important in grasping the significance of performative tactics.[4] An emphasis on ecologies at different (mental, institutional, and environmental) scales has proven pivotal both to developments in the environmental humanities—due, in part, to the influence of Guattari on Isabelle Stengers's cosmopolitics—and to the "turn to practice" in the study of protest cultures.[5] Ecological thought (in the Guattarian sense) and ecological understandings of protest often arise in parallel with one another and frequently have divergent aims or conceptual assumptions.[6] Yet, despite their differences, affinities between protest ecologies and different strands of ecological thought are such that theoretical insight can still be gleaned from the way that activists tinker with specific protest ecologies in practice.

I am drawing attention to these affinities not to revisit conceptual debates about how ecologies of practice can offer new theoretical insight into the workings of sociotechnical assemblages, nor to describe particular protest or media ecologies (especially when these tasks have been taken up so extensively elsewhere). Instead this chapter aims to enrich and complicate understandings of how the open, cosmopolitical approach that has been advocated by strands of ecological thought relates to environmental politics.[7] A deeper examination of long-standing debates within activism about the risks and exclusions inherent in open political configurations can, more specifically, help to contest conceptual distinctions between modes of ethics that are considered to trouble anthropocentric norms and those that are considered to reinscribe them. Such distinctions have become increasingly commonplace in theoretical contexts and need to be disrupted in order to flesh out a clearer understanding of how political stands and interventions can be made without unreflexively reinforcing essentialist values.

A critique of essentialist ethical norms has been integral to the large and ever-expanding body of theoretical work that spans disciplines including science studies, more-than-human geographies, and cultural theory, which has foregrounded the irreducible relationships between human life and more-than-human agencies.[8] From van Dooren's articulation of the irreducible relations among vultures, religious practices, agricultural drugs, and poverty in an Indian context, to Jamie Lorimer's analysis of the thriving communities of invertebrates that have emerged within saproxylic and ruderal ecologies in brownfield sites across the United Kingdom, more-than-human entanglements have been framed as risky, lively, and unpredictable.[9]

As outlined in the introduction, moreover, this work has portrayed the recognition and acknowledgment of multispecies entanglements as vital in generating new forms of ethical responsibility and care that extend beyond the human.[10] A prerequisite for these forms of care to unfold, however, is creating space for particular ecologies to offer up surprises. Normative assumptions about how different species will get along, or about what ecologies are necessary for given species to thrive, are seen as problematic from a cosmopolitical perspective that emphasizes the need for openness and the risk of being surprised by emergent ecologies.[11] This sort of open approach to conservation was illustrated perhaps most famously by Steve Hinchliffe and colleagues, in narrating the stories of voles inhabiting urban waterways near the large UK city of Birmingham.[12]

Voles ordinarily compete unfavorably for resources with rats, as well as being predated by the larger rodents; if conservation is oriented around a biopolitical logic (with decisions made based on the normative characteristics of populations), then the logical conclusion would be to see evidence of high brown rat populations as a signal of a lack of voles. Because voles are a protected species this means that a lack of these animals, in turn, has implications for wildlife conservation initiatives, in terms of whether something is or is not designated a riverbank worthy of protection. A population-based approach, however, did not hold in Birmingham, and researchers found themselves surprised by traces of rats and voles cohabiting within this particular urban environment. The point with this example is that it is dangerous (to again draw on van Dooren) to impose a rigid account of what these ecologies "should be"—and hence of what course of action to take—in advance.

Cosmopolitical approaches to environmental politics demand, therefore, paying close attention to the dynamics of specific ecologies and being open to adapting practice on the basis of the surprises that they offer. Openness in this sense carries a specific meaning, of taking the risk of being affected in ways that might overturn ethical assumptions; this risk, however, is shut down as soon as normative, abstract "solutions" to ecological problems are imposed.[13]

To create space for cosmopolitical engagement, it has thus been seen as vital to refuse to draw on conventional political frameworks—such as rights or social justice—that have a predetermined notion of what convivial relationships between humans and nonhumans might look like.[14] In Haraway's words, "we cannot denounce the world in the name of an ideal world. . . . [D]ecisions must take place somehow in the presence of those who will bear their consequences."[15] When activist campaigns or conservation initiatives draw on normative frameworks, they have, correspondingly, been criticized not only for denying the messiness and liveliness of entanglements but for inadvertently shutting down potential ways of becoming, and future opportunities for flourishing, which could arise from transspecies encounters (such as those that might emerge unexpectedly in brownfield sites or on urban riverbanks).[16]

Though cosmopolitical approaches to more-than-human environments have been central to the environmental humanities in terms of thinking through the logics of conservation practice, as outlined in the previous chapter these principles have also been related to other forms of anticapitalist and environmental praxis: from climate change activism to engagements with the agricultural-industrial complex.[17] The particular way that cosmopolitics has been related to these diverse ethical concerns, however, lies at the root of internal tensions within academic fields that have grappled with these issues, especially between different strands of animal studies.

Tensions are evident in particular in subfields such as critical animal studies (CAS) that have been intensely critical of relational ethics for an alleged failure to take a critical ethical stance.[18] In turn, CAS has been accused of shutting down debate and taking moral solace in the act of denouncing particular human-animal relations.[19] The critical-activist perspective offered by CAS, in other words, has been seen as inhabiting a closed world populated by defined-in-advance animals and humans, who have set ways of engaging, a stance that is decisively not open to having preconceived knowledges (or political actions based on these knowledges) unsettled.[20] Ethical frictions have been especially prominent in debates surrounding Haraway's conception of companion species, an approach that is underpinned by cosmopolitical values and that has also further popularized cosmopolitical thinking across the environmental humanities and animal studies.[21]

As illustrated by the quotations opening this chapter, however, neither CAS nor advocates of relational ethics are as unreflexive as their counterparts would insist. For instance, the insistence on retaining "single-issue" liberation politics in the influential pamphlet *Beasts of Burden*, mentioned above, could sit uneasily with van Dooren's calls to move beyond the sort of ethical abstractions that pit a purified nature against an autonomous humanity. Yet, at the same time as they put forward their argument for animal liberation, the pamphlet's authors recognize the potential limits of animal liberation ideologies in and of themselves. Conversely, while van Dooren is intensely critical of normative ethical assumptions, the significance of his arguments, I suggest, lies not in his emphasis on the entanglements between human and more-than-human worlds, but in his argument that these entanglements are precisely why stands need to be taken.

As these opening quotations suggest, therefore, it is an oversimplification if CAS perspectives are dismissed out of hand on the basis of perceived dogmatism, because key thinkers within CAS have not denied interspecies entanglements in an ontological sense. Instead, critical-activist work has simply called for caution when it comes to relations that could irrevocably damage particularly vulnerable actors or environments. Key theorists within the environmental humanities likewise have recognized that the act of highlighting entanglement is not in itself enough to guard against the transformation of life into resources, with Lorimer warning that "fungible, laissez-faire neoliberal natures and fluid, self-willed ecologies are ontologically not that different" and that an "open-ended ecology of surprises . . . could inadvertently play into the hands of those who would like to see [these ecologies] removed."[22]

Yet, despite their nuances, "critical" perspectives still have distinct differences from "mainstream" animal studies. Warnings about relational ethics, as articulated by CAS, for instance, have implications for nonanthropocentric theories that complicate the way that cosmopolitical approaches have been engaged with across the environmental humanities. As things stand, the uptake and application of open, cosmopolitical ethics has led to repeated reflection that certain forms of relating could be "noninnocent," or that seemingly innocuous relations, and even caring ones, can also serve an instrumental purpose.[23] What still needs further reflection, though, is the way that openness can not only support but perpetuate instrumental relations, and shut down the potential for epistemic surprise and ethical transformation even while giving the impression of doing otherwise.

As outlined in the previous chapter, openness and flat ontologies can often revert to norms that reproduce existing intersecting oppressions. In practice, it is vital to make purposive decisions about which relations need to be excluded in order to create space for less damaging relations to emerge. This line of argument is potentially dangerous, however, in opening worryingly normative questions of which relations to exclude in advance. What I argue in this chapter is that—despite these obvious dangers—the exclusion of particular relations or ways of doing things is not a problem that can be avoided, as even nonintervention and pluralism support a particular materialization of reality at the expense of alternatives. The focus, therefore, needs to be less on avoiding approaches and practices that exclude ways of being, and instead on finding ways to make these exclusions visible in order to foster accountability and create space for these relations to be contested in the future.

It is thus important to revisit the question raised by Susan Leigh Star in the previous chapters, that of cui bono, or who benefits from particular ways of doing things, back to the infrastructures that support openness itself. As argued in the previous chapter, this line of questioning is important in highlighting who bears the brunt of the risks inherent in cosmopolitical modes of relating and—more important—how the asymmetrical burdens of any risks can be lessened.[24] These arguments cannot be made sense of in the abstract, and it is in navigating questions of how to combine a cosmopolitical refusal of anthropocentric norms with a close attunement to the asymmetry of risk inherent in this approach that the instances of activism discussed throughout this chapter offer especial insight. But before I turn to activism itself, it is necessary to elaborate on the existing ways that open, experimental political approaches have been related to political practice.

Openness in Anticapitalist Practice

Although tensions between relational ethics and CAS's commitment to contesting animal exploitation seem to place these perspectives at opposing poles, further dialogue between them is useful in confronting difficult questions about how to navigate instrumental, hierarchical, and risky relations between species. What makes such a dialogue complicated to realize in relation to activist practice is that (as discussed in the previous chapter) related conceptions of openness preexist within activism itself. These subtly different conceptions of openness can pose difficult issues for both CAS and relational approaches to ethics.

Calls for openness have been pivotal to certain strands of anticapitalist activism. As touched on previously, McLibel's significance came—in part—from its relationship to broader protest networks, which were foreshadowed by the campaign. One of the final chapters of John Vidal's 1997 book on McLibel is entitled, somewhat presciently, "And It's Not Just Morris and Steel," reflecting the growing prominence of global anticapitalist protest networks.[25] As touched on in the previous chapter in relation to Indymedia, the global justice movement was often referred to as the "movement of movements" and was composed of a loose network of activist groups based in different global contexts. At its peak the movement gained visibility through these groups working together at specific junctures in order to contest neoliberal processes of globalization.[26] Though, as anti-McDonald's campaigning evidences, these networks didn't come out of nowhere, in 1999 they gained substantial media and academic attention due to high-profile coordinated protests, most famously the Battle of Seattle, in which activists converged in Seattle to protest against the World Trade Organization.[27]

Post-Seattle, high-profile "summit protests" against large transnational institutions, such as the G8, G20, and International Monetary Fund as well as the World Trade Organization itself, became a defining feature of the global justice movement, with the early 2000s seeing increasingly innovative and complex attempts to support protest through creating temporary protest camp infrastructures near the summits themselves.[28] A particularly notable instance of this was Horizone, a camp that was established in the Scottish town of Stirling to meet the 2005 G8 summit at Gleneagles and that had intimate relations with the initiatives discussed in previous chapters. The camp was bound up with the launch of the first (at the time temporary) Indymedia center in the country, for instance, and catering collectives who had been involved with anti-McDonald's campaigning were integral in facilitating food provision.[29]

Though summit protests were some of the most prominent protest events initiated by the global justice movement, these large-scale actions also inspired more local anticapitalist initiatives. Indeed, the realization of protest camps depended on existing activist infrastructures. Space for planning meetings and infrastructure for Horizone were provided by autonomous social centers and community initiatives that had also emerged during this period, which themselves were designed to operate (relatively) autonomously from state control, while remaining nonprofit.[30] These initiatives were prefigurative in their aims and worked to embody values in the present that mirrored future aims of social justice, with these values frequently manifested by having such initiatives operate according to open, consensus-driven, and nonhierarchical principles (though these values were often challenging to prefigure in practice).[31]

Being open to a diversity of opinions, a diversity of members, and a diversity of approaches was central to these protest movements; indeed, one of their core characteristics was again their experimental nature. Activists, for instance, constantly experimented with inventing new political formations and infrastructures to realize nonhierarchical values within environments that were often hostile to these values (as with the media initiatives discussed in the previous chapter).[32] Although autonomous conceptions of *openness* cannot be mapped neatly onto cosmopolitical understandings of the term, due to hailing from a specific political tradition with anarchist roots, they have been sufficiently similar to garner praise from Haraway and Stengers (among others) for embodying the spirit of their arguments.[33]

In light of the affinities between alter-globalization and ecological thought, it is perhaps unsurprising, therefore, that ecological approaches have increasingly been utilized in academic *analyses* of protest to characterize the relationships among different protest actors, the shifting material-semiotic dynamics of protest, and the evolving media arrangements that support activism.[34] Anna Feigenbaum, Fabien Frenzel, and Patrick McCurdy, for instance, draw on Guattari and Stengers to foreground the way seemingly mundane actors within sites such as protest camps—from tents, walls, and media technologies to toilets and saucepans—mediate practice in distinct ways, arguing for the value of an ecological approach in drawing attention "to the importance of movement innovation, non-linear exchanges of knowledges and practices, and the complexity of enmeshed human and non-human networks."[35] Although these protest ecologies enact and prefigure alternative worlds and ways of living, the realities they produce tend to be more fragile than those produced by the expansive sociotechnical networks outlined by Star (as described in the first chapter of this book). Entities such as McDonald's, for instance, can give the realities they forge a degree of ontological stability, due to their capacity to enroll vast numbers of actors. Protest ecologies, in contrast, often have to work tactically within and against the assemblages they are opposing, and have to constantly negotiate the constraints imposed by existing material-semiotic arrangements, even as they experiment with alternative ways of doing things.[36]

The dynamism and experimental qualities of protest ecologies are thus, in part, necessities born of their uneven relationship to the systems they are contesting, but, as illustrated by Indymedia, openness to change and to difference is also an *ethical* principle that underpins the work of these movements.[37] This embrace of openness by autonomous movements, for instance, is encapsulated by Rodrigo Nunes's rallying call: "*Nothing* is what democracy looks like—horizontality is not a model (or a property that can be predicated of things) but a practice. And as a practice, it remains permanently open to the future and to difference."[38] The inherent openness of autonomous protest cultures is thus entangled with an ethical commitment to the values of openness, and it is here that intimate affinities with cosmopolitical approaches can be found.

In drawing attention to these different understandings or indeed contestations of openness—which exist in relation to (critical) animal studies, anti-capitalist praxis (and conceptual frameworks used to grasp this praxis), and more-than-human modes of thought—a messy picture emerges. Viewed in ecological terms, however, this messiness can be understood not as a hindrance to how these forms of politics can be understood but as a means of opening up some productive ways of thinking across different ecological scales and conceptual domains. Open, relational ethics and openness to having particular knowledges unsettled—in order to rethink ways of doing things—seem to be manifested and borne out in practice in certain autonomous protest contexts (such as media initiatives and protest camps), as reflected by both the material implementation of values of openness and the use of cosmopolitical, ecological approaches to analyze these settings. Yet, as hinted at in the previous chapter, when one pays careful attention to the dynamics of openness in activist practice, certain dangers associated with materializing it are also revealed. Openness, in other words, comes with its own exclusions that are often difficult to detect. Examining some of the mundane features of protest camps, however, is useful in making these exclusions visible.

Protest Camps: Composting and Consuming Worlds

Haraway likens cosmopolitical interventions that resist moral absolutes to processes of composting, arguing that "staying with the trouble requires oddkin; that is, we require each other in unexpected collaborations and combinations, in hot compost piles. We become-with each other or not at all."[39] The characterization of compost as a site of experimentation with new ways of "getting on together" adopted a very literal form at the Gleneagles anti-G8 protest camp, in the work of permaculture activist Starhawk, whose presence has particular conceptual resonance due to Stengers, Haraway, and Maria Puig de la Bellacasa's subsequent engagement with her work (in slightly different US contexts).[40] For Stengers, Starhawk's work elucidates that—even though anthropogenic environmental problems have global consequences—it is vital to resist recourse to a universalist politics about how to deal with these issues. Such a politics both reinforces well-worn divides between experts and publics and, in turn, reinscribes geopolitical hierarchies between those who bear the consequence of these problems and those who have the expertise to deal with them. Stengers argues that Starhawk's neopagan empowerment rituals, which evoke witchcraft and magic (and hence might be viewed with derision in other theoretical quarters), offer techniques for connecting activists with the situated environments they are working in. In doing so, rituals offer a means of grasping the specific ways that environmental concerns permeate mental and institutional ecologies: "reclaiming an ecology that gives the situations we confront the power to have us thinking feeling, imagining and not theorizing about them."[41]

To gain a clearer sense of what such empowerment practices might entail, it is helpful to turn to the anti-G8 Horizone protest camp; Starhawk herself contributes a chapter to the 2005 collection of essays *Shut Them Down!* that was compiled and written by activists and academic commentators who were involved with the G8 protests. The entire focus of the chapter was on an important aspect of the protest camps that rarely gets discussed with any seriousness but that provides precisely the sort of compost admired by Stengers, as well as Haraway and Puig de la Bellacasa: toilets. As Paul N. Edwards notes, when the infrastructures that are vital for day-to-day life are running smoothly, they often go unnoticed; it is only when we move to a different context, with different infrastructural arrangements, or if infrastructures break down, that they actively come into view—a principle that enabled Horizone's toilets to play an important prefigurative role in the lives of protest campers.[42] Starhawk's chapter draws attention to the ecological damage that can be wrought by protest camps, and to this end her role at the camp was to facilitate the development of compost toilets and gray water systems:

> For ten days we wallowed in compost toilets and greywater systems—okay, I'm being metaphorical here—we wallowed in discussion of these things, conceiving of ways in which problems might become solutions, waste be transformed to resources, physical structures support directly democratic social structures and people might be encouraged to wash their hands. How many shits does it take to fill a 55 gallon drum, and what is that in liters? What could you do with it afterwards? How many liters of greywater would 5,000 people produce in a week, and where could it go if the clay soil doesn't drain?[43]

What was significant about the construction of sewerage infrastructures was that they prefigured particular ecological values, while also unsettling particular norms attached to acts as mundane as using the toilet, simply by revealing the invisible work that ordinarily allows these acts to function:

> Because the ditches fill up, people have to watch how much water they use. Because we've built compost toilets, we have to actually think about what happens to our shit, and who is going to deal with it. "We're spoiled, normally," a young woman says. "We don't usually have to think about any of this." "It's anarchism in practice," I tell them. "Being self-responsible at a very, very basic level." In that moment, watching the realization dawn on them that water has to go somewhere, and shit has to be dealt with somehow, I feel that all the work and stress of this project has been worth it.[44]

The construction of alternative infrastructures thus necessitated situated construction practices, which were themselves valuable in revealing invisible work and generating questions about the everyday sociotechnical arrangements that naturalize and legitimate this work. Especially important, moreover, was the *way* that toilet construction supported activists' political articulation; insights emerged through situated practices of grappling with compost, via activists crafting collective, but very specific, relations with the local environment. This approach was, therefore, a means of not only enacting connections among mental, collective, and environmental ecologies but posing questions about how these relations were composed in everyday life, in ways that brought questions of whether things could be otherwise to the fore.

As evident in the previous chapter, however, infrastructure creation is never neutral; indeed, it lends itself to social hierarchies, as certain activists will inevitably have existing expertise in undertaking a particular task. A central principle of openness in this context, therefore, is to include additional layers of practices to avoid elitism. Before Horizone, for instance, hierarchical tendencies (tendencies that can even emerge in composting contexts) were deliberately countered through ten days of skill-sharing activities, described by Starhawk, and these location-specific sewerage skill shares were crucial in supporting collective decision-making about the ultimate realization of sanitation infrastructures to avoid top-down processes of design and instead foster situated obligations toward the local environment. Tactics such as skill shares are imperfect and can never hope to entirely combat asymmetries, but they do prevent them from becoming unspoken power differentials that are allowed to flourish unquestioned.

What composting also foregrounds is that even something that seems like an obvious ecological good still entails particular practices being privileged over others, and decisions about which practices to valorize are always driven by ethical imperatives. In van Dooren's terms, stands always have to be taken for some worlds to the exclusion of others. Amid calls to take inspiration from compost as an instance of lively ways of getting along together, it perhaps seems dry to reflect on the practicalities of compost infrastructures, but it is nonetheless important to remember that the realization of such practices can be fraught. In the context of green, anticapitalist activism, developing some form of less environmentally damaging sewerage system might seem uncontentious, but it still necessitates taking a stand; other practices—related to food provision, or direct-action protest tactics—pose more visible and pronounced problems, moreover, and even conflicts over how best to approach things.

"Openness" in Practice: The Incommensurability of Pluralisms

A difficulty that has recurred in autonomous activists' reflections about protest tactics is that avoiding hierarchy is often achieved by allowing a "diversity of tactics" to flourish. *Diversity* in this sense entails individual groups working toward shared aims by using different means, an approach that seems to refuse any sort of purism about which approaches are the right ways of doing things. This approach, therefore, resonates with calls for cosmopolitical openness with regard to environmental relations.

Situations such as protest camps, however, complicate debates about openness and moralism. As in the previous chapter, at times political principles of openness can serve as a platitude that allows hard questions about *which* way of doing things is privileged to go unanswered; certain enactments of openness, in other words, are not always what they seem. Sometimes, for instance, the refusal to criticize particular ways of doing things results in situations where tactics cancel one another out in less-than-obvious ways, as noted by Nunes in his reflections on horizontal forms of political organization at the protests:

> Despite being always discussed, [issues] are almost always solved by some form of application of the principle of diversity of tactics or some sort of

> interpretation of consensus decision-making. Practically, this means that "pacifists" in the "violence" debate are defeated, since their goal is to stop "violence" from happening. . . . This is where the feeling of the debate never actually happening comes from: positions are taken from the start to be absolutes that do not suffer any inflection according to practical, situational contexts—and in fact are absolutely impervious to any debate and can never be changed. Therefore, it becomes a question of one position winning and the other one losing, but this winning/losing can never be acknowledged since making such decisions is bad, because it reduces diversity, and so on.[45]

At Gleneagles these tensions were brought to the fore when local activists (based in the nearby town of Stirling) worked to make the protests meaningful to the local community, through inviting people to the Horizone ecovillage and engaging in dialogue about their work. Connections with local residents, however, were countered by alleged vandalism committed by activists in local towns (with the defacing of transnational chains such as Pizza Hut), which was used by police to foster antagonism between activists and publics.[46] A wariness of hierarchy, and of totalizing decisions, in other words, led to certain actions implicitly dominating praxis by bringing particular relations into being that precluded alternatives. This is not to say that these tactics are wrong in any rigid sense; indeed, it is well documented that liberal appeals for "nonviolence" or "rational discussion" can privilege interlocutors or forms of praxis that maintain the status quo and even silence those who are oppressed.[47] Issues surrounding violence and direct action nonetheless illustrate the importance of refusing to treat pluralism as a solution or guarantee of openness. The aim in drawing attention to the above problems is not to wholly dismiss the principle of a diversity of tactics, but more to foreground the danger of it being seen as an unproblematic solution to hierarchy. The danger of taking solace in this approach to tactics is evident, in this instance, when the material protest actions undertaken by certain activists reinforced rifts between local communities and the protest camp, which not only rendered activist values incomprehensible (if not in opposition to the community's way of life) but reinforced material divides between activists and publics.

My key reason for drawing attention to these issues is that this sort of self-reflexive anticapitalist criticism of "a diversity of practices" can help to enrich debates about openness within nonanthropocentric conceptual contexts by helpfully nuancing distinctions between which practices are seen as totalizing and which are perceived as open. For Haraway, integral to "staying with the trouble"—and indeed integral to relational ethics more broadly—is a refusal to decide what is good or bad in advance, as this has the potential to foreclose alternative ways of being. It is this refusal to decide in advance of the context that, for Haraway, creates space for cosmopolitical openness to having particular (anthropocentric) knowledges unsettled. Though this line of argument has already been subject to critique within CAS, debates about violence and hierarchy highlight a very specific set of tensions associated with calls for openness. These long-standing contentions within activist contexts help to foreground how openness to a diversity of modes of relating can *itself* shut down particular ways of being because the mere existence of certain ways of life or ways of doing things can foreclose others. Sometimes, in other words, the refusal to purposively take a stand can inadvertently allow particular norms, practices, and values to dominate that cannot be combated for fear of essentialism, and it is this problem that has proven especially tricky to negotiate in the context of anticapitalist activism.

Essentialism and Openness within Animal Activism

Debates about openness versus essentialism have particular purchase in the context of animal activism, which has had an uneasy role within contemporary anticapitalist movements, with these divisions paralleling those that exist between CAS and relational approaches (though with a slightly different emphasis). UK Animal activists have argued that a significant disjunction between anticapitalist protest and animal activism emerged in the 1980s, after the thriving anarcho-punk scene—which combined a critique of capitalism with values of animal liberation—became incorporated into more hierarchical Marxist organizations: "Animals were now irrelevant, and if anything eating meat was a badge of the 'ordinary people.' Some 'Vegan police' who had moralistically condemned others for eating meat, now criticised vegetarians for not eating meat: the diet changed but the self-righteous attitude stayed the same. Concern about animals was derided as middle class and liberal."[48] This is just one characterization of divisions between animal activism and other forms of anticapitalist activism (albeit a particularly influential one), and the rifts between these strands of practices was not total. As outlined in previous chapters, initiatives such as anti-McDonald's activism retained strong overlaps with animal activist movements throughout the 1980s and early 1990s. The resurgence of nonhierarchical anticapitalist projects with the rise of the global justice movement also saw vegan collectives such as Veggies and the Anarchist Teapot assume a significant role within protest infrastructures in the United Kingdom, as they were necessary to support transnational initiatives such as Horizone.[49] Tensions between class politics and animal activism were nonetheless prominent, with certain practices promoted by animal activists accused of being hierarchical or elitist, "closed" ethical norms that failed to have their assumptions unsettled by taking account of particular sociocultural contexts (an accusation leveled at veganism in particular). In the late 1990s, for instance, an infamous anonymous pamphlet circulated that argued (among other points):

> In reality the situation is that most people can't afford health foods, can't afford the time or energy to take regular exercise. It's easy to recommend these things, but hard to put into practice if you've got an energy sapping, mind numbing job and/or a couple of kids at your knees nagging you to take them into Mcdonalds which robs you blind as well. Certainly society offers choice, but importantly for a lot of working class people this is just the illusion of choice. For the middle classes, in their blind indifference to the daily suffering and hurt of this society (reinforced neatly by the middle class dominated media and politicians) it is easy to exercise the privilege of economic and cultural choice and attempt to boycott a specific part of the hurt—animal cruelty (even though that is an impossible goal within this society).[50]

Self-reflexive warnings on the part of activists have long emphasized that the values espoused by the movement can be intensely classed, can exclude those outside of the activist community, and can be a barrier to movement building.[51] Internal concerns about the ethical norms and demands of animal activism, in other words, have long mirrored external theoretical criticisms of totalizing moral principles; though specific practices (such as veganism) might be advocated by particular groups of activists, therefore, they have historically *not* offered moral solace as is often assumed by theoretical work.[52]

At the same time as they have engaged in self-reflexive criticisms, animal activists have also sought to contest blanket portrayals of their work and values as totalizing (whether these accusations have been leveled from outside or from within the activist community). Rather than jettisoning the core values of the animal liberation movement outright, on the basis of their apparently totalizing nature, activists have instead worked to nuance and complicate how these values are enacted in practice. To combat the dangers of animal activism becoming a single-issue cause, for instance, particular collectives have emphasized intersections between human and animal labor in order to resituate animal activism within anticapitalist narratives. In the early 2000s, several animal activist pamphlets that explicitly drew connections between human and animal struggles grew to prominence in grassroots circles. The aforementioned *Beasts of Burden* as well as texts such as *Veganism and Social Revolution* worked to articulate the form of animal "histories from below" that have been called for in more recent academic contexts, highlighting shared patterns of labor and processes of value extraction that implicate human and nonhuman alike.[53]

Activists' interventions, moreover, were decisively material-semiotic in their focus, shifting the emphasis from discrete acts of killing to entanglements between labels (such as "animal") and more systematic processes of rendering species "killable" or "exploitable" (to put things in Haraway's terms). In addition, the label "animal" was criticized for functioning as a discursive resource that serves to marginalize certain groups of humans who have historically been "animalized."[54] The approach offered by these texts thus refused to simply extend rights frameworks to "certain privileged others" (the criticism often leveled at critical-activist perspectives in theoretical contexts) but was instead oriented around a recognition of relationships between human and animal labor and called for action precisely on this basis.[55] Indeed, this emphasis on charting intersections has more recently been central to CAS.[56]

In setting out these particular activist histories, my aim is not to uncritically praise animal activist pamphlets: in the debates that emerged in the wake of *Beasts of Burden*, for instance, even sympathetic commentators argued that "the arguments brought forward [in the pamphlet] never manage to confront the inherent moralism of the animal liberation ideology, regardless of whether it can be shown that animal abuse is historically constituted."[57] The pamphlets nonetheless offer a helpful starting point for illustrating how recognition of entanglements between human and animal labor has historically informed praxis. Though this recognition has different ends and is taken to different conclusions than much work in the environmental humanities, it nonetheless provides useful context for developing a more complicated reading of activism that unsettles conceptual assumptions. The remainder of this chapter teases out these provocations by turning to instances of food activism.

In autonomous contexts food has acted as an especially heated site of debate and discussion about which practices prefigure nonhierarchical aims over others and how—to put things in the context of the issues raised in *Beasts of Burden*—animal activism can be enacted in a manner that subverts anthropocentric ways of relating to the world rather than perpetuating closed moralistic (and often inadvertently anthropocentric) norms. A focus on food activism associated with these movements is thus useful in foregrounding how the dangers of moralism, hierarchy, and essentialism are navigated in practice.

Tactics for developing a more complicated food politics are particularly evident in the context of autonomous protest camps. The Anarchist Teapot collective's reflections on the difficulties of creating infrastructures to support Horizone, for instance, detail all manner of debates and decisions that had to be made about which food to source, where to locate it, and how to prepare it. Campsite catering dilemmas ranged from how deal with the "muesli mountains" that resulted from unexpected donations, to how to cater for unpredictable numbers of activists (which led to an overordering of bread and the resultant development of unique culinary creations such as "bread spreads—to put on bread"), or even how to clean 350-liter pans that were almost as large as the activists who washed them up.[58] Echoing the issues that were raised in chapter 1, some of these everyday difficulties, despite appearing trivial, had marked ethico-political implications.

Certain concrete requirements forced compromises: the need to create a workable mass-catering infrastructure in a tight time frame resulted in particular groups having to initiate coordination to ensure—in a very pragmatic sense—that everyone was fed. The scale of the camp, moreover, meant that groups with existing mass-catering experience or access to equipment initially assumed privileged positions. Due to these immediate organizational demands, the initial catering working group was founded by Veggies and the Anarchist Teapot, which "slowly started compiling information, and reaching out to find other mobile kitchens to help cook for the expected 10 000 or whatever random number was being bandied about."[59] As with the compost toilet situation, certain preexisting sociotechnical arrangements (that were difficult for activists to contest) thus resulted in particular groups adopting particular roles; Veggies' position as a registered food business, for instance, meant that they were in a position to navigate the necessary legal requirements in a manner inaccessible to other groups.[60] As with sewerage and media systems, however, a range of techniques were used to prevent the initial asymmetries in skills, knowledges, and experiences from becoming the de facto rule.

Situatedness was again important to prevent particular ways of doing things from being imposed as norms by privileged groups; in this instance the caterers refused to simply use their standard suppliers and worked instead to create more situated connections with the local context. Relationships were carefully developed with local farmers and suppliers in order to source ingredients such as local vegetables, whole foods, and freshly baked bread. To ensure that a nonhierarchical approach informed food preparation itself, connections were also crafted with activist catering collectives based in other contexts to support them in acquiring the resources to act relatively independently rather than dictating what these collectives "should" be doing:

> There were a large number of kitchens, mostly able to cater for 100–300+ people in a neighbourhood: the Belgian Kokkerelen collective; an Irish kitchen combining Bitchen Kitchen and Certain Death Vegan Café; the Scottish Healands kitchen who were already on site when we arrived and who we hadn't heard of beforehand; kitchens from the social centres in Bradford (1in2) and Leeds (Common Place, with some Sheffield people too); Veggies from Nottingham; Why don't you from Newcastle; a kitchen from Lancaster with lots of Danish people for some reason; a Bristol kitchen; a kitchen from Oxford; a kitchen in the Queer Barrio.[61]

As with sanitation, the work of creating infrastructures proved important in creating space for different collectives to acquire the skills necessary for them to work relatively autonomously.

> It took a few days for us all to find our feet—some of the kitchens that came hadn't had what they needed, lots of gas splitters were installed and ditches dug and water needed to be connected up, etc, etc. . . . Everyone seemed to get the hang of it pretty quickly though. . . . I had been worried that we would be too dominant with our huge kitchen and food store, and we would be the "experts" on site, and in some cases I suppose we were, but generally, each kitchen developed its own individual way of doing things and it felt varied and decentralised.[62]

Catering in this autonomous context, therefore, necessitated a careful navigation of values of openness, on the one hand, and the need for concrete action, on the other, and reveals very particular tactics used to aid this negotiation. Embedded within these infrastructures were a number of responsibility practices designed to generate responsibility on the part of activists toward the protest ecologies that they were coconstituting. Responsibility was, for instance, generated through constructing material relations between activists and the specific rural environments they were working in, which in turn fostered mutual obligations among activist collectives, as well as situated responsibilities toward the local environment. As evident in the Anarchist Teapot collective's reflections, however, despite all the work that went into overcoming hierarchical relations, there was still a recognition of asymmetries in resources and expertise between groups, born of mundane constraints. In light of the (often urgent) need to take stands, activists thus face a constant struggle: it is important that fear of hierarchy does not stultify the potential for political action while at the same time ensuring that any informal hierarchies that do emerge are visible and open to contestation.[63]

The difficulties of realizing truly open, nonhierarchical approaches are evident if the focus shifts to food itself, with a decision made before the camp that food would be vegan. While decisions such as this were, in part, due to pragmatic necessity, they again caused debates about openness versus essentialism to rear their heads. Although veganism has frequently been cited as a form of "embodied resistance" that enables activists, or members of subcultures, to contest normative consumption practices, it has (as outlined previously) also been framed as individualistic "ethical lifestylism" with elitist tendencies.[64] From the lens of companion species, and allied nonanthropocentric approaches, these tensions are often resolved through advocating a diversity of consumption practices in a move that instinctively seems to be more open and refuses to let one perspective (or dietary imperative) become an unquestioned norm.[65] Haraway's criticisms of veganism, for example, which have been so heavily contested from a CAS perspective, are due to her wariness of shutting down alternative possibilities of eating.

Yet in the material realization of protest spaces, as argued above, it is dangerous to take solace in a diversity of tactics. As with media initiatives and debates about violence, particular expressions of openness can allow tactics to dominate that foreclose alternatives. Due to this dominance being implicit and linked to preexisting cultural norms, moreover, it is often naturalized and difficult to challenge. These debates help to situate the complexity of eating decisions in activist settings; despite seeming more open, plurality can allow normative patterns of consumption to go unquestioned. In this instance, creating space for a multitude of different ways of eating would implicitly construct "animal issues" as a personal commitment rather than a component of broader attempts to prefigure alternative worlds that intersects with other inequalities.[66]

To go back to the question posed by Lorimer in the introductory section of the chapter, in the context of prefigurative protest sites, the question is really whether openness is actually open, or whether some manifestations of openness lend themselves to uneven distributions of risk and struggle (as occurred in the previous chapter, in relation to media experiments). These questions are especially difficult to answer amid the messy practicalities of autonomous spaces and the role of food within them, which muddy perceptions about which approach to consumption can be constructed as more useful for enacting a stance that stays with the trouble. It is important, therefore, to find a way of moving beyond debates about how best to realize openness as a political value. One way of doing this, I suggest, is to ask slightly different questions that move beyond notions of a "diversity of [food] practices" and instead ask how decisive ethico-political stands within food activism can be enacted in ways that are more accountable to the worlds they bring into being.

Performative Food Activism on the Streets

Further detail about the practice of autonomous food politics can be offered through turning to my own experiences, in a different—but related—site, associated with one of the caterers involved with Horizone: Veggies Catering Campaign. The group was (and indeed is) based in a prominent autonomous social center in the United Kingdom, the Sumac Centre, located in the city of Nottingham. In addition to their work at Horizone, Veggies played an important role in anti-McDonald's campaigning from the mid-1980s, described in the previous chapters.[67] The group not only produced the mass-circulated *What's Wrong with McDonald's?* pamphlet but participated in ongoing protest and fund-raising efforts, with some members even acting as witnesses at the McLibel trial itself.[68] Food has assumed a playful role in the group's activities since their emergence in 1984, when activists presented the president of McDonald's UK with a giant veggie burger, after which they formalized themselves as a permanent catering cooperative.[69] The group did not limit themselves to McDonald's campaigning, however, but acted as "campaign caterers," providing food at prominent peace camps, environmental actions, and animal rights protest marches, as well as local grassroots community events. Though preceding the global justice movement, their work often segued with specific actions that took place under this umbrella in the late 1990s and early 2000s, as with their involvement in Horizone.[70]

Broad-based anti-McDonald's campaigning saw decreasing levels of publicity post-McLibel; though popular cultural critiques of the corporation persisted in the early 2000s (most famously Morgan Spurlock's documentary *Super Size Me*), criticisms often centered on health-related issues—especially in relation to children—rather than the more comprehensive drawing together of environmental, labor, and animal welfare issues that characterized the trial.[71] Both Veggies and members of the local Nottingham animal rights group, however, continued campaigning and distributing pamphlets twice annually on key dates such as the anniversary of McLibel's conclusion and the International Day of Action against McDonald's (which had emerged as a key date for solidarity campaigning during the trial). But with a lack of media focus on the campaign came a lessening in public engagement, and activists increasingly employed more creative methods of campaigning in order to counter this declining attention. From the spring of 2008 onward, free-food giveaways, which involved the distribution of food in prominent spaces in the city center, began to be regularly incorporated into pamphleteering activities. These tactics reached their peak in 2010, when we organized monthly protests that culminated in a multitarget day of action on December 10, which went beyond a McDonald's-specific focus, with five food stalls and five activist-literature stalls spread across the city, in close proximity to busy holiday shopping areas.[72]

Food sharing in public space has been a common tactic in anarchist and autonomous social movements, growing to prominence when Food Not Bombs emerged in Cambridge, Massachusetts, in 1980.[73] After gleaning food that is about to be discarded as waste (or, in some instances, that has already been discarded), local Food Not Bombs chapters collectively prepare meals and share them in public space with homeless communities.[74] These food-sharing tactics are purposively performative; for instance, they actively contest legislation (such as no-loitering, antisocial-behavior, and no-begging laws) designed to exclude homeless populations from central, urban public space.[75] The use of gleaned foods also plays a vital role in articulating a situated anticapitalist critique; by making both homelessness and waste visible, it is possible to bring the contradictory existence of food waste and food poverty into stark relief.[76] Performative food protests are thus material in their effects, actively making a difference to the composition of the city through momentarily disrupting and reworking public space (albeit very briefly).

More, the physical occupation of space is bound tightly with the protests' symbolic significance, in signifying an egalitarian right to the city wherein all populations should be able to "participate in the work and the making of the city and the right to urban life (which is to say the right to be part of the city—to be present, to *be*)."[77] Food Not Bombs uses food, in other words, to articulate connections between issues in order to understand how particular sociotechnical relations legitimize systematic processes of marginalization and exploitation. Through the use of food, however, activists are able to perform this critique in a manner that creates distinct forms of responsibility; the highly visible nature of these protests means they not only are able to physically articulate irreducible connections between issues but involve those who are the most affected by these concerns.

On one level there are thus clear affinities between performative food activism and the tactics central to the McLibel trial, as the giveaways were effectively a public articulation of the issues that were drawn together, in the most painstaking manner, by the McLibel activists. Food giveaways enact these relationships, moreover, through practices that directly involve actors who are enrolled by the networks associated with corporate actors such as McDonald's and, in the process, temporarily disrupt patterns of consumption that are necessary for the smooth functioning of these networks. Yet the potentials of these protests, in terms of fostering responsibility, lie not just in their disruptive affordances but in their dialogical ones. In relation to Food Not Bombs, Drew Robert Winter suggests that the perennial danger of didacticism in food activism can be ameliorated by the performative dimension of food giveaways, with ethics articulated "through doing rather than saying, or showing rather than telling."[78] To build on this point, what is specific to acts of sharing is that they actively construct relations between publics and activists that open specific obligations; in my own documentation of the protests on the activist news site Indymedia, for instance, I noted that "the most important part of the day was the amount of people who approached us and wanted to have long and serious discussions about the reasons behind the protest. It was particularly refreshing to have groups of teenagers approach us and want to talk at length about the importance of considering how what you eat relates to so many other issues."[79] And as I described in a report on the large multitarget protests: "All in all, it was a lot of time and hard work to coordinate, but seemed to be a huge success in allowing us to directly engage with the public, who might not ordinarily have attended events perceived as 'vegan' or 'animal rights related.'"[80] Though I initially wrote these comments as part of Indymedia reports about the protests, which reflected on the value of particular tactics for moving outside conventionally activist spaces and creating opportunities for dialogue, food giveaways also generate new ethico-political requirements beyond these limited reflections. Because they take place in urban space and involve consumers, protests open space to have particular assumptions unsettled, making it difficult, for instance, to maintain notions of passive consumers who only need to have their views of agriculture debunked. Indeed, the protests were characterized by debate and dialogue with regard to how we had represented the issues at stake, and often resulted in us adapting our approach in response to these concerns.

This is not to say that giveaways entirely avoid consecrating uneven relations between the activists sharing food and the people consuming it. Indeed, the challenge of negotiating hierarchy is a well-known problem for Food Not Bombs due to the danger of interpellating those who need food into activists' critique of inequality. Joshua Sbicca, for instance, offers a specific instance of asymmetries emerging in a Food Not Bombs chapter in Orlando, describing how—amid the legislative crackdown on the group in US contexts—activists wanted to maintain the protests while homeless people were not able to risk arrest.[81] Food activism can also be uneasily co-opted into an ordo-liberal agenda through easing the withdrawal of public services. In the UK context that Nottingham Vegan Campaigns was working in, the co-option of food activism was a particular problem, due to the Conservative-led government's contemporaneous launch of a raft of policies under the label "Big Society," which were designed to elide criticisms of its austerity program (which had included sharp welfare cuts) by relying on community food initiatives. Food-related activism thus poses some difficult questions about the potential for radical community initiatives to compensate for neoliberal policy.[82]

Responsibility practices are again important in avoiding these dangers of creating a normative model of what disruptive food politics "looks like": like protest camp infrastructures, food giveaways demand collaboration and hard work from participants, which offers space for constant discussion, and it is in this regard that the materiality of protests is vital to consider. As Feigenbaum argues, "affect and emotion are bound up in object encounters, [and] so too are tactical knowledges and embodied experience."[83] The work that underpins the most mundane infrastructures that support protest, from this perspective, is what prevents things from becoming unspoken norms. The assemblage of material-semiotic relations and hard work it takes to prefigure alternative (if temporary) economies of mutual aid is important in the context of food giveaways. In being so open to contestation, these situated performances can foster a felt understanding of the material constraints of particular local environments. In doing so, they create space to revise practices if new norms and exclusions emerge, in order to avoid some of the pitfalls outlined in the previous chapter.

In addition, as with the skill-share composting workshops engaged in at Gleneagles, food giveaways made (and make) use of a number of more specific tactics to navigate hierarchical problems. Food Not Bombs, for instance, strives to actively involve those they share food with in all aspects of the food-production process (including food-gathering and preparation processes) and explicitly refers to its work as "sharing" food instead of distributing it.[84] We adopted similar tactics in Nottingham, inviting people who had shared food with us in protests to engage in food preparation the following month, and distributed simple, cheap recipes along with food.[85]

Simply creating skill shares, however, does not automatically overcome hierarchical relations due to the highly classed cultural discourse surrounding healthy eating in the United Kingdom (and frequent conflations of veganism with health foods).[86] Indeed, even though working together on practical tasks can open new collective engagements and responsibilities, as Veronica Barassi points out, sometimes the material requirements of particular tasks can, again, result in certain spaces being privileged over others in ways that have political consequences.[87] Mirroring events at Horizone, for instance, hygiene and spatial requirements meant that food preparation for the food giveaways took place in the autonomous social center where Veggies was based. This setting ameliorated certain problems (in enabling the workshops to take place in a space where people were committed to overcoming hierarchical relations), but it was very much a space dominated by activists that could pose barriers for those who were not previously part of this community. Again, responsibility plays an important role: in the case of food giveaways, even though these problems were ultimately unresolved, they were decisively not allowed to congeal as the sort of "sorry, but we have to" logic condemned by Stengers. This logic justifies particular ways of doing things on the basis of preexisting sociotechnical norms (a tactic that is often adopted by corporate actors, as described in chapter 1), without reflecting on whether these norms themselves could be changed. In contrast to this approach, the protest ecologies of food activism were characterized by constant adaptation and the development of new tactics in order to maximize opportunities for participation while still enabling protests to actually take place.

The use of vegan food elucidated particular tensions in both camps and giveaways, simultaneously running the risk of elitism or moralism while also being the tool that supported complex articulations of the relationships between human and animal labor. This way of eating, however, was not approached uncritically. The protest literature outlined previously continuously emphasizes the importance of making connections between the invisible work of human and nonhuman animals, as illustrated by one of the key pamphlets produced in the wake of debates: "When we purchase a food product at the grocery store, we can read the ingredients list and usually tell whether animals were murdered and/or tortured in the production process. But what do we learn of the people who made that product? . . . Were a hundred slaughtered on a picket line for demanding a living wage?"[88] Despite all of their tensions, food giveaways offer a means of addressing the problem of making these messy connections visible, due to the performative way they denaturalize everyday patterns of consumption that occur in specific sites. The regular Nottingham protests, for example, involved activists cooking and distributing veggie burgers created from locally sourced ingredients outside McDonald's, as a counterpoint to the food preparation undertaken inside.

The multitarget protests extended this principle, with a wider range of products that had symbolic ties to the sites where they were distributed. Like Food Not Bombs, therefore, this was a decisively material-semiotic protest act, wherein the symbolism of the food was decisively bound up with the act of disruption. While the broad act of food sharing opened space for dialogue, the nature of this dialogue was further nuanced by the type of food distributed: through vegan food, a point of contrast could be made with the normative practices of consumption occurring in particular sites. This is not to say that vegan food in this context can be seen intrinsically standing in for animal bodies as some form of "absent referent" (to use Carol J. Adams's term), more that the performative distribution of food in specific sites is what gave it disruptive material-semiotic affordances.[89]

In these protests veganism was able to articulate intersections between human and animal labor, for instance, due to being distributed for free in public space and situated as part of broader protest ecologies, where dialogue with activists and protest literature reiterated these relations. Food giveaways, in other words, though not uncontentious, can be used to enact painstaking articulations of issues that are composed with those who are implicated in these relations. These tactics, therefore, offer a means of taking a stand while simultaneously constructing obligations toward that stand.

What is important about food giveaways and protest camp infrastructures is thus twofold: first, these initiatives not only highlight but actively retie the relationships among sociolegal apparatuses, urban planning, and the forms of commercial enrollment described in the previous chapters (as drawn together by McDonald's and similar chains). Whereas the campaigns discussed in the first chapters strove to contest sociotechnical norms by involving those most implicated in the issues at stake, however, giveaways couple this with direct interventions into what Annemarie Mol describes as the "reality effects" of these relations, or the particular infrastructural arrangements that enact and naturalize certain ways of being.[90]

Puig de la Bellacasa helps to encapsulate what is at stake in this approach, calling for "visions that 'cut' differently the shape of a thing" and suggesting that "critical cuts shouldn't merely expose or produce conflict but should also foster caring relations. Such relations . . . maintain and repair a world so that humans and non-humans can live in it as well as possible in a complex life-sustaining web."[91] Though this argument was made in relation to the knowledge politics of science studies, it is equally applicable to the responsibility practices intrinsic to performative food activism. These instances of protest enact tactical interventions by creating particular relationships in order to contest others.

Second, what is important about these forms of protest is that their performative dimension helps to render visible the relations that are being opposed; perhaps more important, this approach creates space for dialogue surrounding particular interventions and the constitutive exclusions they forge. In doing so, performative protest tactics can open political stands to responsibility and future change.

Accounting for Exclusions

What I have sought to elucidate throughout this chapter is that just as more-than-human worlds are composed of messy and irreducible relationships, so too are protest movements. As reiterated by Stengers: "No force is good or bad. It is the assemblage that comes into being when one encounters a force and is affected by it, which demands experimentation and discrimination."[92] Food activism underlines Stengers's point while illustrating that this principle needs to be reflexively extended to *ethical* stands. As argued throughout the chapter, commitments that are routinely condemned for being essentialist or totalizing in theoretical contexts need to be carefully situated. If these ethical stands are seen not as static but as something that is enacted within particular activist contexts—as a component of specific communication ecologies and protest repertoires—then it becomes possible to grasp how even "totalizing" ethical commitments can open space to trouble norms rather than reinscribe them. Situating ethical decisions within broader protest ecologies, in other words, elucidates how the affordances of ethical approaches are themselves shaped through practice.

To build on previous chapters, what the activism discussed here underlines is not the importance of refusing to adopt a staunch ethical commitment but the need to make such stands visible and open to contestation. The insights drawn from food giveaways, in particular, help to offer a sense of where more productive dialogue can emerge that does not perpetuate rifts between critical-activist and relational approaches.

At the same time, the issues discussed in this chapter also offer more profound provocations for relational ethics. Just as practices that are often labeled totalizing can be enacted in provocative and accountable ways, practices that have the appearance of "staying with the trouble" can sometimes foreclose responsibility. As hinted at throughout these first three chapters, in certain environments and political settings, relational ethics can inadvertently shore up hierarchies and inequalities, but does so in ways that are difficult to identify. These problems become particularly profound when it comes to questions of how to be more accountable for the exclusions that are bound up with any course of action.

As Michelle Murphy suggests, an emphasis on entanglement, multiplicity, and complexity can sometimes inadvertently diffuse responsibilities. In the context of very different infrastructures—related to built urban environments—Murphy describes how concepts such as sick building syndrome offered a more holistic understanding of public health that reflected the myriad of factors that could create an environment detrimental to workers. At the same time, this ecological emphasis made it difficult to establish causal links between specific industrial problems and ill health. In this instance, therefore, "ecology helped to materialize other capacities opened by multiplicity" (such as making sick building syndrome an urgent matter of concern), but at the same time "invoking multiplicity could shift the very grounds of causality to a constant uncertainty."[93] In other words, within certain environments an emphasis on irreducible complexity can foreclose responsibility rather than open up space for new responsibilities.

If related to broader questions of political intervention, Murphy's argument reveals the danger of calls to "stay with the trouble" becoming reduced to a platitude, a danger that becomes especially pronounced if this sentiment simply serves as a recognition that no position is innocent. Understanding Haraway's slogan as a call to recognize complexity and maintain openness can make it difficult to locate coordinates for intervention, or determine how to contest relations that impact on the most vulnerable.[94] To echo Murphy, this emphasis can also actively shore up oppressive relations due to the particular ways it can make responsibility so difficult to realize.

It is this paradox of relationality that is addressed in the final chapters, which explore how the foreclosure of approaches that seem totalizing can inadvertently dismiss productive forms of ethical intervention. These interventions, I argue, are often far messier and more productively ambivalent than they first appear, and insights can be drawn even from approaches that are routinely criticized in theoretical contexts (for being abstract or sentimental, for instance). Conversely, as elucidated throughout this chapter, dismissing these perspectives out of hand—for being overly totalizing or essentialist—is misleading in failing to situate these perspectives in the very environments that give them their affordances. But, more than failing to do justice to the work accomplished by particular forms of intervention, sidelining "critical" approaches out of hand is ethically and politically dangerous. As I argue throughout the rest of the book, valorizing certain modes of ethics—even situated relational ones—at the expense of others can inadvertently shore up normative hierarchies of expertise based on proximity to what is being cared for. In paying attention to the work accomplished by more contentious forms of intervention, I aim in the rest of the book, therefore, to not just recuperate critical approaches but, in doing so, push for responsibility for the exclusions that are constituted by relationality itself.

4 Hierarchies of Care

To promote care in our world we cannot throw out critical standpoints with the bathwater of corrosive critique.
—Maria Puig de la Bellacasa, "Matters of Care in Technoscience"

World Day for Laboratory Animals was instituted in 1979 and has been a catalyst for the movement to end the suffering of animals in laboratories around the world and their replacement with advanced scientific non-animal techniques. The suffering of millions of animals all over the world is commemorated on every continent.
—World Day for Laboratory Animals, mission statement

April 24 is World Day for Laboratory Animals and, as the activist mission statement above suggests, is traditionally marked by demonstrations, protest marches, and awareness-raising events, with the aim of commemorating animals—past and present—used in experimental research.[1] In 2007, for instance, UK activists marked the day with a national protest march, which was held in the city of Oxford due to the recent construction of a primate laboratory at Oxford University. The laboratory itself had been the focus of a controversial campaign by the animal rights group SPEAK since proposals for the laboratory were announced in March 2004.

I was attending the event with one of the food-activism projects discussed in the previous chapter but had no personal experience of protests related to animal research, and beforehand I felt nervous. Due to repeated mass-media references to the militancy of antivivisection groups, in addition to internal criticisms of militant approaches to activism that stemmed from within the animal activist community, I was concerned about the potential dynamics of the march.[2] I found, however, a fundamental disjunction between the representations of antivivisection groups that were a familiar part of the media landscape and the descriptions of these groups given by the friends and colleagues I had worked with in relation to food activism, which made me sense that the issue was messier than it seemed—a feeling that was borne out by my experiences on the day itself.

On one level, in my initial impressions, the march did seem to reflect stereotypes about animal activism that existed not just in the mass media but in theoretical contexts. Commemorating animals could be seen as precisely the form of sentimentalism that has been used to trivialize activist concerns from the birth of the antivivisection movement or, latterly, to exclude them from debates about laboratory ethics.[3] The march also seemed to be precisely the sort of activism that has been troubled by theoretical work, particularly work that has criticized rights language for its essentialism and extension of rights to "certain privileged others" (in this instance macaque monkeys) while shoring up human privilege by situating people as ventriloquists for "nature."[4] The protest march, in other words, was seemingly born of precisely the sort of ethical commitments—or mode of "corrosive critique," to put things in Maria Puig de la Bellacasa's terms—that have been treated with such wariness in theoretical contexts.

Indeed, my own experience of the march, at least superficially, bore out characterizations of activism as polemical, emotionally invested, and rights oriented. The premarch gathering point was a playing field next to a parking lot, which was populated with "fringe science" stalls and vegan food outlets.[5] After winding noisily through the city, the march itself culminated with speeches in which people visibly showed emotion. One speech even described the launch of a campaign that encouraged emotional identification with a macaque at Oxford named Felix.[6] Yet, though orienting a campaign around Felix seems to bear out characterizations of animal activism as paternalistic and sentimental, the situation was more complicated than this.[7] Felix had actually been named by researchers and—prior to the "Felix Campaign" itself—used as an emblem in mass-media narratives to demonstrate the care given to nonhuman primates at Oxford, when they were used within deep-brain-stimulation research.[8] By appealing to very different constructions of Felix's subjectivity, then, activists were working within existing, mediated terms of reference, while also attempting to challenge them.[9]

The clash between my preconceptions and experiences, coupled with the way the activists themselves engaged with media representations of the issue, underlines the need to pay attention to how debates about activism are constituted by the mainstream media. My own initial perceptions of these particular antivivisection groups, for instance, were bound up with media narratives about animal rights activism, in ways that hint at some of the issues activists face. However, in this chapter I am not just interested in questions of representation and mediation in and of themselves but in the conceptual significance of these issues in relation to care ethics. What is revealed when turning to some of the ways the Felix Campaign was mediated is not just the integral role of distinctions between "legitimate" and "illegitimate" knowledge but the ways these distinctions are often entwined with particular modes of care.

In drawing attention to the relationships between care and knowledge politics, in this instance, my aim is to complicate a theoretical emphasis on care as something that unfolds through proximal encounters and entanglements between bodies. An emphasis on proximity has been central to work that has explored modes of generating ethical responsibility between human and nonhuman bodies and has proven especially influential in the context of laboratory research. Rather than exploring the ethical potentials that bodily relations and entanglements bring into being, however, this chapter underlines the importance of addressing what they *foreclose*.

The forms of care ethics discussed throughout the chapter are intended to create space for the agency of those most affected by a particular issue or situation. Here, however, I illustrate that in contexts where those affected are not human, an emphasis on proximity can lend itself to modes of care ethics that are practiced by experts, which are often inaccessible to critical perspectives and thus consecrate epistemological or political inequalities. What I am not seeking to do, in drawing attention to these dynamics, is uncritically valorize all forms of "marginal" knowledge: as contemporary concern with post truth politics or climate skepticism has foregrounded, to do so without attending to the political motivations and contexts that inform such dissenting knowledge is a dangerous move.[10] Despite these issues it is nonetheless important to understand how and why—in certain contexts—forms of ethics and epistemological approaches that are routinely presented as *unsettling* normative sociotechnical arrangements can sometimes *reinforce* them through rendering criticism impossible.

While this chapter delineates hierarchies of care that emerged in relation to specific primate research controversies, the following two chapters bring home the stakes of these hierarchies by examining the fraught politics that surround two particular ethical frameworks—suffering and sentimentality—

that are routinely foreclosed by approaches that emphasize bodily entanglements and encounters. In doing so, the chapters are designed to build up a multilayered picture of practices that are excluded by ethics born of particular forms of entanglement and to explicate the stakes of these exclusions. This chapter establishes the coordinates for these broader concerns by focusing on the Felix Campaign: a case where I suggest that hierarchies of care, based on proximity, played a constitutive role.

Mediated Activism

The contemporary antivivisection movement is bound up with distinct histories of representation. From a social movement studies perspective, it has been argued that media depictions of animal rights groups—and antivivisection activism in particular—have been especially hostile within the United States, western Europe, and Australia.[11] As I discuss in this chapter, present-day narratives have grown out of preexisting discourses and frames associated with antivivisectionism since the late nineteenth century. Protest against animal research has long been subject to very visible and public contestation, from the Victorian origins of the contemporary animal rights movement, and these representational histories are critical in grasping the constraints activists have to work within and against.[12] The historical significance, contemporary prominence, and transnational influence of activist groups originating in the United Kingdom have led to these groups having a central role in debates about media representations of animal activism.[13] Yet while activists and social movement research have foregrounded bias against activism, scientists who use animals in their research, in contrast, have argued that the media are biased against *their* work.[14] In light of these issues, an emerging body of work has illustrated how this sense of media bias—that is shared by both activists and researchers—has resulted in each group carefully articulating their beliefs, values, and practices in relation to their respective perceptions of public attitudes toward their work.[15] Media representations, in other words, are an important backdrop to some of the tactics engaged in by the different parties involved in these debates.

As described above, my initial concern with media representations of activists' ethical commitments emerged from auto-ethnographic reflection: in particular, the disjuncture between my perceptions (based on popular cultural portrayals of animal activism) and my lived experiences. In the rest of the chapter, I work to situate these experiences by engaging in a documentary analysis of texts relevant to debates about animal research. A range of materials are drawn on to elucidate how the controversies (which surrounded both the laboratory and activist campaigns against it) unfolded in the media, including in newspaper articles focused on these debates, a documentary (*Monkeys, Rats and Me*) broadcast at the controversy's height, and reframings of these mass-media narratives that were engaged in by key antagonists within the online network that grew out of the controversy.[16] This online "issue network" drew together a number of actors who were central to the debate's constitution in mass-media contexts, including SPEAK, the Safer Medicines Campaign (an anti-animal-research lobby group spearheaded by medical professionals who were drawn on regularly by SPEAK to support its criticisms), Pro-Test (a campaign set up directly to combat SPEAK, which was established by an Oxford teenager but used by researchers as a platform for their arguments), and Understanding Animal Research (a pro-animal-research advocacy group drawn on frequently to bolster Pro-Test's arguments).[17]

Through a close analysis of these materials, what emerges as particularly significant in conceptual terms is the role of contrasting understandings of care in fostering distinctions between irrational and rational publics. Hope has been attached to speculative care ethics, as a means of addressing the obligations that particular human communities should have toward specific multispecies entanglements. Yet even the core proponents of care ethics have argued that caring itself can still create priorities, hierarchies, and exclusions.[18] In order to enable certain ways of life to flourish, the needs of any entities that clash with the object of care may have to be neglected, marginalized, or even denied. The prominence of care within discourses surrounding SPEAK helps to further enrich conceptual arguments about care ethics, by drawing attention to the way that particular appeals to care can determine whose knowledge, beliefs, and practices are made to matter and whose are excluded from consideration.

Although it is important not to homogenize the diverse bodies of work that have focused on the ethical and epistemological value of care, loosely speaking, two particular lines of argument have proven especially important in attempts to support ethical engagements with more-than-human worlds. First, there has been an emphasis on the everyday practices involved in care work (in contexts that include laboratory research and conservation), with bodily relations seen to offer a source of insight and ongoing ethical responsibility on the part of humans toward animals.[19] The second dimension of care pertains to work that has emphasized care's capacity to support new forms of knowledge politics, which are attentive to divergent ways of knowing. Though bodily care and care as knowledge politics often overlap, both have proven important in developing ways of thinking and acting that decenter possessive-individualist humanism. It is thus useful to grasp the specific conceptual implications of each mode of care, and for this reason the following chapter will explore the significance of care as a form of embodied ethical engagement, while this chapter will explore what Puig de la Bellacasa describes as the "speculative ethics" of care.[20]

Care as Knowledge Politics

The problem, brought into stark relief by Bruno Latour's much-cited essay "Why Has Critique Run Out of Steam?," is that cultural theory's desire to debunk normative constructions of the world by foregrounding their implication in power relations has not only mirrored conspiracy theories and antiscience skepticism but actively lent itself to the support of such perspectives: "Entire Ph.D. programs are still running to make sure that good American kids are learning the hard way that facts are made up, that there is no such thing as natural, unmediated, unbiased access to truth, that we are always prisoners of language, that we always speak from a particular standpoint, and so on, while dangerous extremists are using the very same argument of social construction to destroy hard-won evidence that could save our lives."[21] In other words, when confronted with issues such as anthropogenic climate change that hold implications for diverse forms of life, should the focus of cultural theory remain on debunking facts in order to reveal power relations?[22] Latour advocates instead that "facts" would be better understood as "concerns," which theory should aim to "add" reality to rather than "subtract" reality from.[23] What this means in practice is that theoretical work should aim to draw attention to the messy "web of associations" that create particular realities in order "to detect how many participants are gathered in a thing to make it exist and to maintain its existence."[24] Such a move, Latour notes, could itself easily be seen as subtracting reality from things, but this interpretation holds only if we accept the conceptual bifurcation of facts from falsities, which (he alleges) characterizes critical thought. A more productive approach, Latour continues, would be to understand the act of criticism not as debunking facts as false constructions but as *assembling* the relations that lie behind particular realities and enable them to exist. As a concept, then, "matters of concern" pushes for a richer, multilayered understanding of the relations that constitute lived reality.

Puig de la Bellacasa's vital rejoinder to Latour advances his arguments by stressing the value of assembling the associations and actors that lie behind realities, and elucidates how his approach is valuable in foregrounding the distinct critical, ethical, and political commitments that underpin particular constructions: "The purpose of showing how things are assembled is not to dismantle things, nor undermine the reality of matters of fact with critical suspicion about the powerful (human) interests they might reflect and convey. Instead, to exhibit the concerns that attach and hold together matters of fact is to enrich and affirm their reality by adding further articulations."[25] The articulations central to matters of concern might include, for instance, the political contexts supported by particular facts; the ethical frameworks they create or perpetuate; and the sociotechnical arrangements they underpin. This approach, crucially, also emphasizes the role of nonhumans in mediating particular assemblages, striving to give voice to the important (but often unstable or at least complicated) way in which things are contingent on particular actors translating concerns into sociotechnical realities.

Yet, while intensely sympathetic to Latour's stance, Puig de la Bellacasa productively complicates the concept of matters of concern by calling instead for "matters of care." What her call for care works to do is "add layers of concern" to how issues are staged within cultural theory; these new layers "are not necessarily incompatible with [the] mediating purpose" of Latour's matters of concern, "but would represent and promote additional attachments." In particular, enriching matters of concern by drawing on feminist visions of care accentuates questions that have been central to feminist science studies, with its emphasis on invisible work and interrogation of who benefits from this work, rearticulating "questions about *who* will do the work of care, as well as *how* to do it and for *whom*."[26] Care, crucially, also "connotes attention and worry for those who can be harmed by an assemblage but whose voices are less valued, as are their concerns and need for care."[27] The final dimension of care that Puig de la Bellacasa calls for relates to specific demands on researchers that pertain to "the researcher's own cares" and, more specifically, "what are we encouraging caring for?" How, in other words, does the articulation of an issue by researchers "intervene in how a matter of fact/concern is perceived"?[28]

An emphasis on care, in other words, is important in reintroducing (perhaps unfashionable) critical questions about how to respond to irreducibly entangled worlds. What is important, moreover, is that care offers a means of doing this that avoids an "abuse of notions of power, used as causal explanations 'coming out of the deep dark below' to undermine what others present as facts."[29] Implicit in Latour's line of argument, therefore, are normative claims about how knowledge should be produced that Puig de la Bellacasa reconstructs as decisively ethical questions. The normative dimension of their respective arguments, however, means that both Puig de la Bellacasa's and Latour's work open particular questions about, to echo the refrain of previous chapters, what could be *excluded* if normative understandings of care itself were to come into being.

Caring in the "Wrong Way"?

As touched on in previous chapters, the exclusions bound up with particular epistemic norms are made explicit by Puig de la Bellacasa herself; a particular point of criticism she levels at matters of concern relates to the implications of Latour's arguments for activism. Citing a passage in which he derides the criticism of SUV users by "angry environmentalists," Puig de la Bellacasa warns that "respect for concerns and the call for care become arguments to moderate a critical standpoint. The kind of standpoint that tends to produce divergences and oppositional knowledges based on attachments to particular visions, and indeed that sometimes presents its positions as non-negotiable—what Latour has named 'fundamentalism.' This dialogue thus also exhibits mistrust regarding minoritarian and radical ways of politicizing things that tend to focus on exposing relations of power and exclusion—here the angry environmentalist."[30] Though it is important to note (as Puig de la Bellacasa does) that Latour's aim is against a particular mode of critique (rather than against activism), the use of "angry environmentalists" as an avatar for what he criticizes has distinct ethico-political implications. To go back to Puig de la Bellacasa's aforementioned assertion: "To support a feminist vision of care that engages with persistent forms of exclusion, power and domination in science and technology . . . we cannot throw out critical standpoints with the bathwater of corrosive critique."[31] This reflection is crucial in recognizing the ambivalence of care ethics. On the one hand, it points to the value of treating diverse perspectives with care—even contentious ones. On the other hand, an imperative to care could bring with it normative judgments that exclude critical viewpoints, especially if certain groups' caring stance is (for instance) perceived as angry or so focused on caring abstractly about a particular cause that it is insufficiently caring toward other interlocutors in the debate. This suspicion of critical viewpoints for caring in the "wrong way" might only be a latent theme of Latour's arguments, but the stakes of this inference come to the fore in the mediated terrain that both activists and researchers were forced to negotiate in controversies surrounding the Felix Campaign.

The media narratives surrounding SPEAK illustrate how particular perspectives can become delegitimized if they are framed as caring "irrationally." The dominant media framing of the controversy at Oxford, for instance, portrayed a "battle" where "scientists" and "fanatics" clashed on the "frontline."[32] A prominent researcher at Oxford, Tipu Aziz, featured in sixteen newspaper interviews at the peak of the controversy and was the main spokesperson for researchers in the documentary *Monkeys, Rats and Me*.[33] His central claim was that "every medical therapy that exists today has come out of animal research," an argument taken up by other sources (including Pro-Test and a number of articles) to frame activists as "misanthropic" and "fanatical."[34]

Aziz's assertions were part of a broader discourse that constructed animal research as a straightforward tool that could be used with certainty, which was articulated as a direct response to emotive framings of animal research as unnecessary torture that were expressed across SPEAK's protest slogans, literature, and web resources, and reinforced by images of primates with electrodes in their brains (on leaflets, posters, videos, and their online image gallery).[35] The certainty of animal research's utility in medical research (as articulated by researchers), in other words, was leveraged to combat activists' ethical assertions, with the effect of portraying these concerns as the product of misplaced sentiment that led to activists valuing animal life above human life. These narratives seem, on a basic level, to be part of a long-standing framing of activists by the medical establishment that has roots in the Victorian period, something made explicit in the documentary.

Causal Narratives: Monkeys, Rats and Me

Monkeys, Rats and Me is particularly helpful in delineating the contours of the debate, with researchers' narratives framed explicitly in relation to activism. Within the documentary, for instance, Aziz is quoted using his own deep-brain-stimulation (DBS) research to performatively debunk activist criticisms: "You take someone who's bound to a wheelchair and unable to move or look after themselves, and then suddenly they have surgery based on conclusions drawn from my research and suddenly they're up again, they're restored to human dignity."[36] This assertion supplies the foundation for the story that underpins the documentary, which frames the debate in a way that perpetuates long-standing discourses of activists as privileging animal life above human life.

The documentary is structured around the intertwined stories of Felix and a thirteen-year-old child who has a neurological disorder that profoundly affects his mobility. Aziz's description of the causal relationships between his research and medical progress is, therefore, neutrally presented as a synecdoche for this overarching story. The documentary begins, for instance, with an opening speech that states, "Without an operation developed on monkeys, Sean faces a life of terrible disability" and culminates by depicting a procedure on Felix immediately before a similar operation is conducted on the young patient. The documentary thus offers a visceral representation of the sort of tropes that have been central to the vivisection debate since early twentieth-century physicians engaged in heated letter-writing and pamphleteering campaigns (as epitomized by Joseph Lister's famous refrain "Shall we save a rabbit and allow a man to die?").[37] In *Monkeys, Rats and Me*, this argument is underlined when the disembodied narrator questions the child himself, whose initial ethical stance against animal research is portrayed as having shifted after his operation. After initially voicing concern about animal research in the early scenes of the documentary, the child is questioned again about his feelings after he has surgery and gains some mobility, and he nods in assent when asked if he now believes primate research is justified.

The ethical closure offered by this story is reinforced by the concluding discussion, which again infers misplaced sentimentality on the part of activists, as underlined by director Jonathan Wishart's voice-over: "Although in my heart I'm queasy about seeing what we do to animals in the name of science, in my head I think that the research I've seen is justified." In addition to the discomforting way a child's experiences were utilized as a framing device for an engaging story, therefore, the documentary's arc instrumentalizes this story still further in order to lend support to a particular ethical argument.[38]

However, even as *Monkeys, Rats and Me* firmly aligned activist concerns with latent misanthropy, it also offered openings for activists. Wishart's portrayal of particular technical claims as wholly certain provided anchor points for activists to develop critical counternarratives and avoid the sorts of appeals to emotion that could be used to label activist beliefs as irrational. Causal claims made about research throughout the documentary, as dramatized through entwined human and nonhuman primate stories, ultimately created space for criticism, especially the inference that primate research necessarily leads *directly* to a positive outcome for human patients.[39]

These counternarratives revolved in particular around the role of DBS, the technique that was central to the documentary. To grasp the significance of these debates, therefore, it is useful to offer some further context about DBS research. The technique involves implanting electrodes into the brain and has been experimented with in relation to a number of physiological and psychological disorders.[40] In science studies the development of DBS has itself been framed as an entangled process, born out of an amalgam of commercial, legislative, scientific, and ethical concerns.[41] In tracing these narratives, John Gardner foregrounds how the procedure was initially developed in the United States by "learning in practice" with patients suffering from chronic pain, and argues that it was tightly bound up with the commercial development of medical devices that could be used to administer DBS.[42] The regulation of medical devices in 1976, and ethical concerns over the relationship between commercial and research imperatives in relation to DBS more specifically, however, put a halt to working with human patients until Alim Louis Benabid's Parkinson's research in the 1980s.[43] What was pivotal to the construction of DBS as a treatment for Parkinson's was the development of a primate model on which the necessary research could be conducted to quantify the treatment's success and meet the new regulatory requirements.[44]

Gardner highlights that finding an animal model was complicated by the fact that nonhuman primates do not suffer from Parkinson's, but—as discussed in depth within the neuroscience literature—this changed with the discovery of the neurotoxin 1-methyl-4-phenyl-1,2,3,6-tetrahydropyridine (MPTP) by William Langston.[45] In addition to its role in research, this drug proved to be especially politically significant and played a central role in the tactical uncertainty work activists engaged in. It is thus necessary to give a brief overview of key neuroscience literature that evaluates the use of MPTP, which gives a sense of the drug's function in both enabling comparative studies of Parkinson's and creating sources of uncertainty about the translatability of such studies. When MPTP was initially administered to nonhuman primates, they displayed the symptoms of Parkinson's; the drug induced what was to become known as *parkinsonism*.[46] It was through this process that the necessary comparisons between human and nonhuman primates could be constructed in order to develop the evidence base for quantifying, regulating, and legitimating both DBS as a treatment and the use of medical devices to administer it; MPTP, therefore, was pivotal to securing both the research and the commercial interests bound up with DBS.[47]

The drug MPTP, however, did not resolve all of the issues that had inhibited the construction of DBS as a treatment before 1987. Despite a broad consensus on the part of researchers that primates treated with MPTP have generated new knowledge, uncertainties surrounding individual primate bodies are also emphasized in the literature.[48] Researchers have argued, for instance, that the use of a neurotoxin to alter the chemistry of the brain does not unproblematically translate to the degenerative development of Parkinson's in humans.[49] A number of different factors have to be taken into account to ensure interexperimental parity, with debate over which species are most appropriate (that have persisted since early studies), and a range of different methods being used to quantify success.[50] On an intra-experimental level, moreover, to negotiate uncertainties that arise due to differences in the administration of MPTP and its variable effects on different animals, it is necessary to be attuned to the behaviors of individual primates. Marina E. Emborg contends, for instance, that "early symptoms may be difficult to spot without appropriate training and knowledge of both normal monkey behaviour and the characteristic signs of PD [Parkinson's disease]. It is critical to closely monitor the animals (e.g., body weight, feces characteristics) and to provide supportive care as needed."[51]

The above neurological developments are important to sketch out as they informed the narratives that emerged among researchers and activists within *Monkeys, Rats and Me*; in newspaper articles; and in online contexts. Within the neuroscience literature, while moments of friction and bodily resistance that complicated the translation of primate research were referred to, they were also framed as routine uncertainties of the sort intrinsic to any form of research. From the perspective of activists, however, these tensions offered openings to unpack the narratives articulated in mainstream media contexts, most notably within *Monkeys, Rats and Me*.

Uncertainty Tactics

Drawing attention to technical difficulties in the neuroscience literature was a central tactic. The website of SPEAK, for instance, included a direct riposte to claims made in *Monkeys, Rats and Me*, an article ("Parkinson's—the Truth") in which SPEAK highlighted the use of MPTP and emphasized the variability in how Parkinson's symptoms manifest themselves in macaques who have been administered the drug, as well as differences between neurotoxin-inflicted parkinsonism and Parkinson's disease.[52]

The certainty of the claims about DBS that were made within the documentary gave antivivisection groups several different lines of attack: SPEAK, for instance, drew on peer-reviewed articles based on research that had previously taken place at Oxford that—at least seemingly—did not fit the narrative of research leading to direct medical advancement (for example, "brain damage in monkeys increases their fear of toy snakes").[53] Activists also drew on an article published by Claude Reiss (a key figure in the Safer Medicines Campaign) that questioned the inference that the technique had been pioneered at Oxford: "I do not question the skills of Prof Aziz as a neurosurgeon practicing DBS. His claim to be the discoverer of DBS is however at odds with the fact that DBS was discovered by chance by a team led by LA Benabid in Grenoble (France)."[54] A slightly different tactic was adopted by the Safer Medicines Campaign itself, which submitted a formal complaint to the BBC regarding the strength of the assertion that *all* progress was contingent on animal research (as well as their own omission from the debate, after Wishart did not use material he had filmed after interviewing the group's spokesperson, on the basis that he felt it was at odds with mainstream scientific opinion).[55]

Central to these debates is a tension between activist attempts to portray uncertainties as speaking to intrinsic problems and the researchers' framing of these uncertainties as routine parts of the research process. *Uncertainty* has a specific meaning in this context; Susan Leigh Star identifies several distinct forms of what she describes as "uncertainty-work" within the laboratory, processes that are not purposive obfuscation but an "inextricable part of scientific work organization" as researchers negotiate everyday technical, taxonomic, political, and diagnostic uncertainties—to ensure they do not disrupt the "global certainty" of matters on which there is a broad consensus.[56]

Star's framework is helpful in understanding the dynamics of these controversies about primate research, but here uncertainty work emerged not in the context of producing research findings but as a feature of the mediated representation of debates that surrounded these findings. The engagements with uncertainty she describes were a key characteristic of debates surrounding the Oxford laboratory, where narratives of the "global certainty" of animal research acted as the horizon for debates within the mass media.[57] In *Monkeys, Rats and Me*, for instance, this certainty was appealed to by researchers and pro-research lobbyists to counter activist framings of their work as unethical and to present a united front against (what they described as) activist "junk science."[58]

While researchers appealed to a "global narrative" about the certainty of primate research, as described above, activists focused on what they perceived to be technical uncertainties where particular laboratory tools failed to yield the desired data (which they highlighted within the neuroscience literature). This emphasis on local uncertainties was essentially a form of tactical uncertainty work, designed to support activists' own narrative of the global uncertainty of animal research. The uncertainties identified by activists, moreover, often pertained to the actions of individual monkeys, something that has long been seen as posing technical challenges, not least by Star herself, who draws on monkey behavior as a key instance of how local uncertainties can manifest themselves. In her historical discussion of the tensions that arise when mapping nonhuman primate brains onto humans, for instance, she argues not only that the "mischievous" behavior of monkeys created technical uncertainties in early neurological studies but that issues of "control and specificity" persist in contemporary research.[59]

However, rather than straightforwardly adding legitimacy to activists' arguments, appeals to neuroscience often intensified the fraught debates that surrounded the laboratory. The tactic of focusing on uncertainties—which might seem everyday for researchers and insufficient to trouble global certainties—has been used by antivivisection groups since their Victorian origins.[60] Such tactics have, therefore, long been a source of contestation within the animal research community.

In the case of debates surrounding the Oxford laboratory, activist narratives were dismissed as "junk science" (with Reiss labeled an "ideologically-committed" individual who was "completely isolated" from the mainstream scientific community).[61] Substantial time in *Monkeys, Rats and Me* was also dedicated to exploring differences between applied and experimental research. In the documentary, the perspective of a researcher who conducted a neurological procedure on a rat, for instance, was framed as a redress to antivivisectionist representations of such work as "useless," with the researcher claiming, "For hundreds of years it's basic experiments like this that have given doctors the foundation on which to build new therapies."[62] This discussion of applied versus experimental research was also taken up by Pro-Test, which dedicated substantial space to picking apart activist claims and—mirroring SPEAK's tactics—linked to more authoritative advocacy groups (in this case Understanding Animal Research) to provide evidence of the value of animal research.[63]

More broadly, activists' focus on tensions that researchers see as routine uncertainties has resulted in frustration about perceived misrepresentation in the media (though, as depictions of SPEAK illustrate, in practice activist frames rarely gain media purchase). Nicole Nelson, for instance, describes how techniques that were regularly deployed to control particular research narratives *within* the scientific setting were regularly found "ineffectual" outside of it, which "made researchers feel as though they had little control." In particular, researchers saw themselves as being "in kind of a long-scale war with PETA and the rest of the animal rights movement over the rationale and the validity of animal models at all."[64]

Contestation over certainty and validity, however, was only one dimension of these debates, as what was at stake was not just clashing epistemic commitments but what it means to *care* for nonhuman animals.

The Expertise to Care

What is especially evocative about the controversies discussed in this chapter is that they draw attention to how knowledge claims about research can become intimately bound together with articulations of care. Media portrayals of these debates were not characterized by the sort of "modest witnessing" described by Donna Haraway, wherein truth claims rest on the construction of particular ways of knowing as objective, neutral, and emotionally disengaged.[65] Instead, expertise was consecrated through appeals to bodily care, fostered through researchers' everyday entanglements with laboratory animals.

The documentary, for instance, foregrounded concern on the part of researchers that activists were representing their work as uncaring. Researchers' narratives were explicitly articulated as a response to perceptions of what society in general and activists in particular were felt to believe, in order to debunk notions of uncaring scientists.[66] In articulating research (and researchers) as caring, therefore, media portrayals of the debates worked to undermine not only activist knowledge claims but also their *care* claims. These arguments hinged on distinct notions of bodily expertise on the part of researchers. Scientists were routinely framed as caring due to their knowledge of animal needs, which was acquired from their bodily engagements with animals in the laboratory. This construction of care as a form of bodily expertise is significant in implicitly excluding those without this proximity from caring "legitimately" or, at least, from knowing what "good care" looks like; this narrative thus had distinct ethical implications.

Just as primate research has already been discussed extensively in relation to local uncertainties, it has also been the focus of a large body of theoretical work that is concerned with the capacity of affective encounters to enable animals to impose their own "requirements" on researchers.[67] This line of argument is developed more depth in the next chapter, but the crux of this perspective is that if researchers are open to being affected by the animals they work with, then this offers the foundation for more responsive relations of care toward research partners that exceed the bounds of formal ethical frameworks.[68] It is not enough, from this perspective, for experiments to adhere to formal ethical requirements, as this neglects the particular needs of particular actors, which might not be accommodated by one-size-fits-all standards.

The narrative constructed by *Monkeys, Rats and Me* initially seems poles apart from theoretical work that has advocated a situated approach to care, with researchers repeatedly describing rats and monkeys as "lower animals." Yet within the documentary repeated appeals to species hierarchies were juxtaposed with assertions that researchers' "felt experience" gave them insight into animal well-being.[69] As with Felix, the rat featured in the documentary had a name—Philip—which was given by a researcher who stated, "I quite like to have a friendship with my rats, because I think of them not as a number but as my friends because they're helping me discover things that could help medical research."[70] In handling him, she notes the signs that indicate good health: that his fur is "sleek" and his eyes are "not watery," and that he appears happy as he is not displaying signs of distress. Throughout the encounter there are also repeated inferences that the researcher is gaining direction from Philip; before picking him up, for instance, she points out he may not "want to come."[71] Philip's anaesthetization is even preceded by her explaining that she would stop if she felt the procedure caused him pain, thus depicting him as a research partner whose capacity to signify distress could bring the process to a halt.

The human-animal relations that were articulated by the documentary, therefore, seem to make visible the affective labor involved in constructing human-primate comparisons (as also highlighted within the neuroscience literature by Emborg).[72] Subsequent depictions of Felix's amenability to handling, the emphasis on the "months of training" dedicated to him as an individual, and the reassurance that "pain was not part of the process" built further on this framing of the animals as research partners whose acquiescence and well-being were integral to the experiments.[73]

In this media environment, however, the act of making everyday care practices visible did not serve to disrupt ethical and epistemic hierarchies (a quality often attributed to this mode of bodily care).[74] Rather than opening space for sustained ethical reflection about animals' needs, the representation of these practices was utilized instead to construct hierarchies of care between researchers and activists that, in turn, consecrated epistemic hierarchies. Within these narratives, not only were protestors positioned as irrational (in valuing animal life above humans), but any ethical issues they did have were depicted as unfounded due to their ignorance of animals' "real" needs. What can be drawn out by focusing on these examples, therefore, is the more instrumental side of care work, as it is materialized not just in the laboratory but as a productive material-semiotic configuration within the wider media environment.

The representation of care in the documentary can also be read against the grain in productive ways. While care and attention are explicitly foregrounded, a closer reading of the documentary elucidates the noninnocence of these modes of care: Felix's months of training aim to make him a suitable model for examining the effects of DBS on motor skills and are necessary to ensure he can be handled easily as an experimental subject before making him "parkinsonian."[75] After the emphasis on the care given to Philip, and his capacity to exert agency, this is similarly foreclosed via his anaesthetization, the implantation of electrodes, and his eventual euthanizing. Resonating with research that has explored the more instrumental, violent capacities of care, here care relations are portrayed as valuable in ensuring animal agency does *not* disrupt predetermined goals.[76] The overarching causal narrative that is articulated by the documentary's story, however, ultimately obscures these more instrumental dimensions of care by juxtaposing them with appeals to researchers' caring expertise and using an overarching human-interest story to offer utilitarian solace.

In line with Karen Barad's argument that ethics, epistemology, and ontology are entangled, the material arrangements that enabled animals to be treated as research subjects were legitimated (and reinforced) by utilitarian ethical narratives that, in turn, shaped how the primates were conceived as actors. As outlined above, the term *lower animals* was a regular motif that—when juxtaposed with representations of bodily engagement—ultimately reinforced utilitarian concepts of research as being in the greater good (e.g., by the narrator as well as researchers within *Monkeys, Rats and Me*). Material concerns, such as animals' inability to consent or ethical incommensurability, were, conversely, evoked as the rationale for this lower status.[77] Indeed, Peter Singer loomed large over the debate, with his appearance in *Monkeys, Rats and Me* garnering media attention in its own right, due to the apparent support he lends to the research (deeming it to be in the "greater good").[78]

From the opposite perspective, even amid their criticism of laboratory science, suffering still served as the dominant narrative for activists and legitimated their own self-representation as "the voice for the animals." The material-semiotic constraints related to the particular contexts and spatial arrangements bound up with the controversy, therefore, ultimately offered limited possibilities for how the different parties involved could engage with the issue. Care itself, moreover, assumed a particularly instrumental role, because the capacity to care in the right way was repeatedly associated with forms of somatic knowledge that were accessible only to experts.

Caring for Felix

If a superficial reading is taken of the above events, the dominant trend within activists' self-representations appears to be characterizing nonhuman primates as being of comparable moral standing to humans and as being too different for research to be effective. Mass-media depictions of animal research, in contrast to activists' narratives, seem to emphasize moral differences and physiological similarities between human and nonhuman primates. Yet, though the activist and mass-media narratives were, to an extent, making contrasting arguments, these arguments were ultimately more complicated and far messier than this simplistic interpretation. As nonhuman primate research was articulated in different ways by the different actors involved in debates surrounding the laboratory, what emerged was not a clear-cut dichotomy between appeals to emotion on the part of activists and appeals to scientific progress on the part of researchers. Despite taking different stands, these opposing narratives about primate research were ultimately diffracted through one another; in working to contest opposing claims, each narrative borrowed the terms of reference, vocabulary, and frames more commonly associated with the other side of the debate. Narratives about research did not just focus on epistemic claims but emphasized bodily care in ways that worked to delegitimize activists' ethical stance, while activists tactically drew on peer-reviewed articles to make epistemic claims about uncertainties in neuroscience research.

But there is more going on here than different configurations of care and knowledge. What is vital to take into account is the histories, settings, and structural inequalities that inform these configurations. It is important to recognize how narrow constructions of care and expertise offer significant constraints—for both activists *and* researchers—regarding how ethical positions can be articulated. These concerns are brought to the fore in relation to scientists' concern with articulating their own stance in relation to perceived public opinion (what Pru Hobson-West and Ashley Davies have termed "societal sentience"), on the one hand, and activists' attempts to gain legitimacy by grounding their arguments in relation to the neuroscience literature, on the other—both of which are tactics with long histories, in relation to animal research and more broadly. While foregrounding shared—if unlike—constraints on researchers and activists is important, however, it is also important to recognize that the interlocutors in these sorts of debates are rarely working on even ground.

Standard epistemic hierarchies between experts and publics have, in the case of debates surrounding animal research, long been bound up with hierarchies of care: researchers are positioned as holding expertise not just to *know* about animals' needs but to *care* for them. As is made explicit in *Monkeys, Rats and Me*, activists' lack of proximal engagement with the animals they claim to speak for can be used to cast doubt on the legitimacy of their ethical assertions as well as their expertise. What is especially apparent in these debates is how care is used to distinguish between expert knowledge—gained from working directly with animals—and abstract care grounded in "Disneyfied" conceptions of animals, which is presented as having no place in rational debate.[79]

The significance of these hierarchies of care deserves further attention. Prioritizing proximal modes of engagement and bodily entanglements effectively also prioritizes internal, sympathetic criticism while foreclosing concerns perceived as abstract, external, and inexpert. Bracketing aside the specifics of the particular controversies discussed in this chapter, to focus instead on their dynamics: an emphasis on proximity means that it is not just the capacity to have a "legitimate" opinion that is placed out of reach of particular communities but also the capacity to care legitimately. The danger opened up by this prioritization, in other words, is that voices who are critical of a given set of relations could be permanently excluded from consideration for failing to participate in the very relations they perceive as harmful.

In a contemporary sociopolitical context where critique of expertise has been linked with post truth, the blurring of emotion and knowledge claims in activists' tactical appeals to uncertainty needs to be treated with caution. At the same time, caution also needs to be taken when dismissing these tactics. Latour argues that it is cultural theory that has given skeptical groups the tools to deconstruct scientific claims (a theme revisited in recent debates about post truth and science and technology studies).[80] As the long history of antivivisection activism shows, however, appeal to uncertainty is not a contemporary phenomenon. More, this approach has been employed by a diverse range of movements, which might include the ideologically motivated climate skeptics pointed to by Latour but could also encompass lay publics who could offer valuable expertise that is not recognized as such or patient groups who seek to be involved in debates that have broad sociocultural implications.[81] With such a diverse range of groups employing these tactics, it is important not to uncritically valorize or dismiss activist appeals to uncertainty. What is important is to ensure that the possibility for dissenting knowledge is not rendered structurally impossible.

What the controversies discussed in this chapter illustrate, then, is that care can have an intensely ambivalent role that holds wider theoretical implications. The particular hierarchies of care explored here complicate some of the existing ways that care is often understood in theoretical contexts: as something that troubles anthropocentric norms, creates space for marginalized voices to be heard, and—in cosmopolitical terms—ensures decisions are made in ways that open obligations to those they most affect.[82] If care instead shores up epistemic hierarchies, then this can have the consequence of deciding in advance who has a stake in a particular issue, in ways that reinforce the delegitimization not just of "angry environmentalists" but of other communities who refuse to engage in relations they believe are damaging.

While recognition of the ambivalence of care is already central to Puig de la Bellacasa's work, my concern is that—when appeals to care are engaged with more broadly in theoretical contexts—its constitutive exclusions are often simply *acknowledged* and seen as an instance of the noninnocence of any form of relation. These exclusions, however, need to be engaged with in more sustained ways, as recognizing noninnocence does not get to the root of things: that, in certain contexts, care is not intrinsically troubling to epistemological, ethical, or indeed anthropocentric hierarchies but can shore up inequalities. This argument is fleshed out in the final chapters, where I work to more actively recuperate the cultural politics of concepts that frequently lie at odds with forms of care ethics pushed for in contemporary theory but that nonetheless do productive work in activist contexts: suffering and sentimentality.

5 Charismatic Suffering

An affective mode of engagement is not necessarily positive, for either humans or other species.
—Jamie Lorimer, *Wildlife in the Anthropocene*

APRIL 2012
HISTORICAL DAY! Photos of beagle puppies passed over Green Hill's barbed wire fence flashed all over the world during a protest at the facility. The now unstoppable protest grew and came under a mass media spotlight. Public support ballooned and support came from associations, individuals and politicians. Green Hill's days were numbered.
—Oppose B&K Universal, timeline of protest at the Green Hill beagle-breeding facility

As described in the above extract, on April 30, 2012, a set of images of laboratory beagles circulated through the social network accounts and websites of grassroots animal activist groups across Europe before filtering into the mainstream media.[1] Such images have traditions as long as the contemporary antivivisection movement itself. As with Felix, the macaque monkey at the center of the campaigns discussed in the previous chapter, particular animals are often treated as icons for animal rights protest.[2] Similarly, particular laboratories or breeding facilities are often used as focusing points, drawing together critical narratives about animal research. What was distinct about these images of beagles, however, was that they documented an action that had already taken place: the release of thirty dogs from Green Hill, a breeding facility in Brescia, Italy, that supplied animals for research purposes. The images, of activists handing beagles to one another while traversing barbed-wire fences, have since been seen as key in drawing wider attention to the Green

Hill campaign, creating public pressure that led to further investigation and resulted in the eventual police seizure of the facility's animals.[3] Eventually all of its 2,639 beagles were adopted by members of the public, and in 2015 several Green Hill employees were prosecuted on the basis of veterinary evidence.[4]

The vivid depictions of small dogs being carefully passed between the hands of activists, over fences and through barbed wire, are characteristic of the type of "image events" that Kevin DeLuca describes as a staple of contemporary transnational environmental activism.[5] Central to these sorts of images are activists placing their bodies in vulnerable positions to draw attention to causes, something only enhanced by the forms of "citizen camera-witnessing" that have emerged with the rise of social media.[6] Embodied vulnerability, in this instance, seemed to be shared across species, with both activist and canine bodies depicted in precarious positions during the release (albeit asymmetrical ones). In a sense these activists were thus using tactics common to protest movements, placing their bodies in dangerous positions in order to construct a shared vulnerability between their own corporeality and their cause.[7]

DeLuca makes this point about embodied vulnerability in relation to dramatic image events related to antiwhaling and antilogging activism. This tactic assumes especial conceptual significance in the context of animal activism, however, due to the long-standing criticisms leveled at activist groups because of their "distance" from the messy realities of direct engagement with animals in favor of sentimentalized and abstract ideals about how animals should be treated (as outlined in the previous chapter). A simple reading of the Green Hill images would place them in contrast with the distancing approach taken in the Felix Campaign, because of the way the beagle release brought human and animal bodies together in very specific ways. Representations of the release draw together two opposing cultural engagements with animal bodies in order to construct a particular, charismatic articulation of bodily vulnerability: the careful handling of beagles echoed everyday engagements with domestic companions and contrasted starkly with the backdrop of barbed wire and high fences, with this contrast legitimizing interventions that sought to remove the beagles from this "unnatural" setting.

What was distinct about Green Hill, moreover, was that the construction of specific forms of human-dog relations as unnatural was not just an activist narrative but (as will be discussed in more depth later in the chapter) central to particular legislative reforms that formed the basis for the ultimate prosecution of Green Hill employees.[8] These affective images were useful not only in gaining public support, therefore, but in focusing expert attention toward the specific, situated needs of the beagles in question. Charges of abuse did not relate to the dogs being "beaten or otherwise physically harmed."[9] Instead, the problem was that "they were unable to express the ethological behaviour of their species—particularly in relation to the dogs having access to spaces where they were able to socialize—as evidenced through a series of etho-anomalies found."[10]

The success of the campaign meant that activist appeals to beagle charisma, coupled with their use of social media, also proved valuable in renegotiating activists' identity and actions as morally justified. Discomfort at the political purchase gained by these images was pronounced across the European animal research community, with concern that this tactic of appealing to beagles' charismatic vulnerability had vindicated antivivisectionist sentiment.[11] Attempts to frame activist arguments as irrational (akin to those delineated previously), however, were more difficult to articulate, in light of the eventual verdict and the veterinary evidence that gave rise to it.[12] By using social media to circulate affective imagery, therefore, activists were able to both gain visibility for their cause and create space for self-representation.[13]

As Zizi Papacharissi has argued, despite all of the shortcomings of social media, when evocative political images are circulated online, this can allow individuals to "assemble around media and platforms that invite affective attunement, support affective investment, and propagate affectively charged expression."[14] Although there is certainly no direct causal relationship between social change and social media, then, particular platforms can facilitate the emergence of counterpublics around particular issues, in ways that are politically significant. Papacharissi's line of argument was borne out by the Green Hill campaign, where the actions undertaken by activist groups were even cited in the final verdict as what had generated public pressure, prompted formal investigations of the facility, and ultimately led to prosecution.[15] The online circulation of affective images, in other words, played an important role in intervening in a facility that was ultimately deemed "unbearable for [beagles'] ethological characteristics."[16]

Positive appraisals of activist tactics on the basis of these political successes, however, obscure some of the tensions associated with their work, which—from the perspective of Donna Haraway, for instance—means such tactics are far less troubling to anthropocentrism than they might initially appear. Despite superficial affinities with strands of theoretical work that have focused on the bodily, the tactic of circulating image events on social media clashes with conceptual attempts to draw attention to the latent anthropocentrism of nonhuman charisma.[17] While particular forms of nonhuman charisma might lend themselves to the affordances of social media, and be important in enabling images to go viral, instrumental uses of charisma also run the risk of reinstalling anthropocentric species hierarchies, where animals close to humans are afforded attention (or indeed protection) at the expense of others. Although the Green Hill images depicted a coming together of different bodies, they thus ultimately appeal to a very different affective logic—grounded in sympathy toward charismatic animals—than theoretical work that has focused on bodily encounters as a site of ethical responsibility.[18]

In turning to Green Hill, this chapter picks up on questions raised previously about hierarchies of care, and the implications of these hierarchies. As I go on to argue, there has been a tendency to position embodied care as more troubling to anthropocentrism than normative frameworks of animal rights and liberation. By extension, long-standing frameworks—grounded in appeals to animal suffering, for instance—are routinely excluded from theoretical consideration due to being associated with an anthropocentric humanism that is impossible to accommodate within situated, relational ethics. Building on the previous chapter, in focusing on controversies surrounding Green Hill, here I elucidate how conceptual distinctions between forms of care do not just have implications for activism but hold broader significance for how the ethics of multispecies relations are conceived.

As I foreground throughout the chapter, when turning to the longer histories of beagle use in experimental research, instrumental forms of bodily care are revealed that trouble some of the broader claims that have been made about its capacity to unsettle ethical and epistemological norms.[19] Conversely, when focusing on Green Hill specifically, the role of suffering is revealed to be more complicated than it might seem on the surface: in these campaigns nonhuman charisma was not used just to ground abstract rights claims or perpetuate stereotypes of cute, friendly animals. Instead, this apparently anthropocentric focus on suffering was the means through which public and legal attention was directed toward the everyday bodily needs of beagles (and how these needs were not being met) in ways that fostered situated ethical obligations toward these specific animals. If appeals to beagles' charismatic suffering are taken seriously, therefore, then this can muddy sharp dichotomies between abstract, rights-based care ethics and situated care.

Throughout this chapter I take a lead from the messiness surrounding care ethics to argue that tactics that might *appear* to be fundamentally untroubling to anthropocentric ways of thinking and acting (at least, from a theoretical perspective) can nonetheless open up unexpected ethical challenges to anthropocentrism. These challenges, however, are in danger of being foreclosed if particular relational modes of ethics are valorized at their expense. Taking the work of seemingly totalizing ethical commitments seriously is thus significant not just in its implications for practice. Engaging with the messy and productive work achieved by these frameworks can also reveal particular forms of theoretical normativity that have emerged in relation to broader attempts to depart from anthropocentric ways of conceiving of and engaging with the world.

Nothing Is What Trouble Looks Like

Before addressing questions of what it *means* to trouble anthropocentrism, it is important to clarify what this distinct use of *trouble* is referring to. In describing her own fieldwork into dietary habits, Emily Yates-Doerr suggests that "staying with the trouble . . . need not itself be troublesome."[20] This argument is evocative; Yates-Doerr's point is made in relation to doing fieldwork and marks a critically important refusal to force participants' responses to fit ready-made narratives (in this instance about constructions of obesity). But, though this suggestion was intended as an aside, the fact that it can be made at all is significant. *Trouble* in the sense intended by Yates-Doerr refers to Haraway's conception of the term, and marks a cosmopolitical refusal to rush to hasty diagnoses of problems that lead to the imposition of totalizing ethical solutions. What does it mean, however, if it becomes possible for an approach that is intended to refuse easy solutions to be described as "untroublesome"? Haraway herself has drawn on numerous slogans to encapsulate her politics, from "cyborgs for earthly survival" to the more recent "make kin not babies," both of which are designed to trouble preexisting ethical norms; a slightly different type of slogan could be crafted, however, if inspiration is taken from the activist movements whose work has been at the heart of previous chapters: "nothing is what trouble looks like."[21]

What this slogan articulates is a resistance to any set of academic practices or ideas becoming a normative model for being "troubling." Staying with the trouble is intended to resist the "moral solace" of totalizing ethical imperatives. It is dangerous, therefore, if solace is found in a particular conception of trouble itself. Just as in activist media experiments, protest camps, and food activism, where rigid models of openness can lead to inadvertent political exclusions and acts of symbolic violence, it is worrying if a particular knowledge politics is seen as intrinsically generative of trouble. What needs to be guarded against, in other words, is the tendency for particular modes of thinking or acting to become normative approaches for guaranteeing trouble, because even conceptions of trouble can carry exclusions.

While Yates-Doerr's suggestion was made in an informal context, and points to some important ethical considerations for fieldwork, this notion of "untroubling trouble" is nonetheless evocative and offers a helpful lens through which to critically engage with the politics of entanglement. Throughout this book, for instance, I have drawn attention to a number of instances of "troubling normativity" that have emerged in different disciplinary contexts. The introduction and first chapters pointed to the danger of valorizing relationality itself. Related conceptual approaches for navigating entangled terrains, such as particular models of experimentation and openness, have also been problematized, by drawing attention to the forms of intervention that are foreclosed by these approaches (as in the instances of activism foregrounded in the second and third chapters). As argued in the previous chapter, even care itself can be enacted in normative ways that can disperse ethical responsibilities at the same time as criticizing existing forms of activism (the danger Maria Puig de la Bellacasa warns of). Here, however, I focus on dangers associated with a particular set of norms surrounding *embodied* care ethics and the ways these arguments are bound up with questions of nonhuman charisma. It is the particular relationships between care and charisma that are foregrounded and complicated by activist appeals to the suffering of charismatic animals within social media campaigns.

Before turning to the beagle-release images as illustrative of some of the tensions that have emerged in relation to charisma, therefore, it is necessary to flesh out a deeper understanding of how embodied approaches to care have been seen to trouble anthropocentric modes of thought and action, and why they might not be as troubling (at least in this regard) as they appear initially.

Care and Charisma

As described in the previous chapter, care has been positioned as an important epistemological orientation; conceiving of particular assemblages as matters of care, for instance, draws attention not only to the disparate relations and actors that are knotted together but to the affects, interests, inequalities, and harms that are bound up with these relations. A different, but overlapping, understanding of care has also proven important for conceiving of ethical engagements with more-than-human actors, one that focuses instead on the capacity of situated bodily engagements to create space for mutual understanding and (to draw again on Haraway's term) response-ability. As Isabelle Stengers argues in "The Cosmopolitical Proposal," to avoid imposing technocratic or anthropocentric logics that predetermine in advance what ethical encounters should look like, it is necessary to create techniques that ensure decisions are made in the presence of those they most affect. It is thus important, from a cosmopolitical perspective, to create room for those being engaged with to speak back and signify their needs in some way. Facilitating processes of speaking back, however, is especially difficult when those involved are not human, not just because of a lack of shared language but due to uneven structural relations that—as Matei Candea points out—might make it difficult for certain nonhumans to impose their requirements on humans.[22]

Care, as enacted through bodily engagements, has been seen as a particularly important technique for creating space for animals to articulate their requirements, especially in contexts that are riven with instrumental relations, such as the animal laboratory. A growing body of work within animal studies, and animal geographies in particular, has argued that creating space for somatic encounters can enable researchers to take an interest in the specific needs of specific animals, which (in Vinciane Despret's terms) can "render-capable" researchers of accomplishing new tasks, generate new knowledge, and create new ethical obligations.[23] It is this form of situated care, which emerges through technicians' everyday care work for animals, that Haraway draws attention to within *When Species Meet* as a means of opening up ongoing ethical response-abilities toward laboratory animals. Beth Greenhough and Emma Roe have extended this argument still further, arguing that these "somatic sensibilities" can create a felt understanding, which extends notions of consent beyond the human.[24]

This focus on the bodily, for Greenhough and Roe, is seen to address two problems. First, it overcomes the limitations of "informed consent" for particularly vulnerable communities of human patients. The assumption that written consent equates with ethics, they suggest, is complicated by issues that range from a lack of understanding of what is being consented to to subsequent discomfort experienced by patients after giving initial consent. From this perspective, everyday care practices and bodily engagements with patients can draw attention to ongoing ethical questions and responsibilities that do not end with the signing of a consent form. Paralleling the ethical problems associated with gaining consent from human patients, the adherence to existing regulatory frameworks for animal ethics in the laboratory is often used as a form of ethical closure; as long as ethics guidelines are adhered to, then this is where responsibility begins and ends. For Greenhough and Roe, somatic sensibilities are a means of fostering more ongoing responsibilities on the part of researchers toward those they work with, by drawing attention to the needs, requirements, and resistance of research subjects, which might not be captured by existing guidelines. Embodied care, in this sense, carries significant ethical and epistemological freight; as Greenhough and Roe argue: "Experimental research frameworks (working with either human and/or nonhuman subjects) could be reconfigured and enhanced through an emphasis on care-relating, consent, and cooperation articulated through bodily communications."[25]

Care, in other words, is positioned as something that holds the capacity to trouble ethical norms and assumptions, including anthropocentric notions of moral closure. Existing research on the histories of laboratory animal ethics, however, poses some difficult questions about whether situated, embodied care is as troubling as it might appear, questions that are brought to the fore when considering the differential relations fostered by nonhuman charisma.

Jamie Lorimer delineates three particular modes of nonhuman charisma: ecological (which relates to characteristics of organisms that are relatively stable, such as physical traits), corporeal (which relates to bodily engagements), and aesthetic (which relates not just to an entity's visual appearance but to emotional impact). A humpback whale apprehended during a whale-watching cruise, for instance, might be perceived as ecologically charismatic due to its size and dramatic relationship to its environment. The corporeal relationship with the whale, though not as proximal as more everyday charismatic encounters (such as playing tug-of-war with a beagle), would still foster a sense of the humpback's charisma through the awe experienced as vast bodies playfully splash into the water to rock small tourist boats. The drama of the scene is likely, therefore, to create a distinct affective charge and emotional response, which could in part be a reason—Lorimer suggests—for the level of attention and conservation energies afforded to particularly charismatic entities. Though he defines three modes of charisma, this is intended to be heuristic as these different forms of charisma are often wholly entangled (as evidenced by the intimate connections among whale size, the effects of whale bodies on the tourist experience, and the related possibilities for affective engagement, for example).[26]

Despite a degree of consistency in how certain organisms are apprehended, it is important to note that charisma is not a fixed property but pertains to "the features of a particular organism that configure its perception by humans and subsequent evaluation. It is a relational property contingent upon the perceiver and the context."[27] Yet although charisma is not static, charisma—and the affective responses bound up with it—can maintain a relatively stable form over time and space if the sociotechnical assemblages through which the charismatic entity at stake is perceived are (relatively) similar.[28] Anders Blok, for instance, describes how particular constructions of whales as endangered and in need of protection (constructions that are central to high-profile transnational environmental nongovernmental organizations) have been furthered by Japanese environmental groups' encouragement of ecological tourism. Creating opportunities for whale charisma to manifest itself has proven especially valuable for Japanese activists in a situation where tensions have historically existed between the situated understandings of whales that have existed in a Japanese context and the "global-scale assemblages" that regulate whaling activities.[29] If the lens of nonhuman charisma is related to this context, then it is possible to see how the particular set of relations fostered by whale watching enabled charisma to be manifested in a specific way, which allowed parallel conceptions, understandings, and affective responses toward whales to traverse "sociospatial boundaries."[30]

Appealing to animals with particularly charismatic traits, therefore, has utility for environmental politics, but charisma also poses danger and can result in a particular biopolitics of conservation wherein the desire to preserve the life of a species can result in the promotion of fixed ontologies. What Lorimer describes as the "ontological choreography" that informs much contemporary conservation practice, for instance, embodies a conception of "biodiversity as biopolitics" that secures "life at the scale of the population" and works to preserve the environment that has traditionally enabled the species at stake to flourish.[31] This sort of "species logic" creates entities who can be seen as "killable" for the greater good, which include any other (more common or less charismatic) species that pose a threat to the endangered species, or that are seen as expendable sacrifices (in support of tasks such as surrogate breeding).[32] The overarching consequence of this mode of conservation, for Lorimer, is that it ensures that "landscapes get frozen in the past" in a "reactive management of extinction."[33]

To revisit the issues described at the start of chapter 3, biopolitical forms of conservation management enact a purified nature that not only unhelpfully masks existing transspecies entanglements but wholly forecloses the possibility of epistemic surprise that might lead to ethical reevaluation of how intervention can and should occur. Indeed, Lorimer's arguments are borne out by some of the stories touched on previously, which elucidate the dangers of purification narratives, such as Thom van Dooren's evocative descriptions of the relationships among religious practices, livestock, and *Gyps* vultures in India, or Steve Hinchliffe and colleagues' accounts of symbiotic relationships between water voles and rats near the UK city of Birmingham.[34] The point, with these examples, is that it is dangerous (to again draw on van Dooren) to impose a rigid account of what these ecologies should be, as the preservation of rigid ecologies on the basis of preserving particularly charismatic species neglects the lively forms of *sympoiesis* that occur when different species make worlds together in unexpected ways.[35] Yet, although charisma might be a risky foundation for an environmental politics, especially for organisms who might be sacrificed for the sake of their more charismatic relatives, those who possess charisma are likely to be in a beneficial position.

The advantages that charisma is alleged to carry in terms of *conservation*, however, are not necessarily shared in other environments. As I have argued in previous work with Gregory Hollin, for instance, entities in a number of other settings can be understood as having distinct, charismatic qualities that emerge within particular sociotechnical assemblages. We originally made this argument in the context of health-care settings, in relation to disorders that had gathered significant attention within neuroscience research. In relating nonhuman charisma to this context, our aim was not just to elucidate the applicability of Lorimer's arguments to these environments but to illustrate "the potential non-innocence of charisma for charismatic organisms themselves. Analyses of healthcare have long detailed—whether through processes of medicalisation or subjectification . . .—the ambivalence of falling under the gaze of medical professionals. If medical attention is, at times, unwanted then charisma may be likewise."[36] Our arguments in relation to the clinic, I suggest, can also be extended to laboratory science, where often it is the distinct charisma of particular animals that has led to their consolidation as "model" experimental organisms, due to their capacity to be cared for easily.

The utility of relating Lorimer's concept of charisma to health-care settings is in its capacity to grasp the relationship between particular sociotechnical assemblages and sets of affective relations, which—together—make the traits of certain clinical conditions visible. As with whale watching, given sociotechnical assemblages produce particular affective relations that structure the manifestation of charisma, and can make particular traits visible in ways that go beyond conventional laboratory apparatuses. The stability of the charisma attached to particular entities in the laboratory setting, moreover, becomes a critical area of concern when it comes to laboratory animals (and activism surrounding particular species). Nonhuman charisma, in laboratory settings, complicates narratives that have emphasized the potentials of somatic care to foster ongoing response-ability and open space for epistemological transformation.

Tensions surrounding the relationship between care and charisma become visible when turning to a very different beagle photograph from the Green Hill images, this time one that speaks to some important histories related to laboratory beagles. Although dogs, including beagles, had been used in experimental research throughout the early twentieth century, the first concerted effort at their standardization emerged in the 1950s as part of an experiment to monitor the effects of radiation on a living population. The first large-scale experimental beagle colony was associated with a long-term experiment at the University of California, Davis (1951–1986) that was funded by the Manhattan Project, subsequently labeled "life-span" radiation experiments. The Davis colony eventually became one of five across the United States that utilized 5,389 dogs.[37] In research about the colony that I undertook with Hollin, what emerged as particularly striking for us was the way that beagle charisma was entangled with the dogs' capacity to be cared for easily, which was in turn bound up with their suitability as experimental animals.

These relationships are illustrated by a particular image from the colony, which featured researchers measuring beagles and appeared in an early academic article that offered guidance on standardization.[38] The image itself depicts an animal perched on a weighing table, being handled by a group of researchers. What is notable is that the dog is not restrained in any way, with no muzzle or even a lead to restrain her. The lack of constraints in this image is notable in light of relational care theories, which have argued for the importance of learning from the bodily needs of animals in order to create space for them to impose their requirements on researchers. From a perspective grounded in encounter-based ethics, this image could be seen as an indication that researchers had met ongoing requirements toward their research partners, with a lack of "objection" interpreted as a sign of contentment. Indeed, Hollin and I have elucidated how this line of conceptual argument resonates with claims made by researchers, who aimed to learn from beagles' bodily behavior in order to care for them more effectively and "establish an ecologic balance or utopic environment that would minimize the over-all effects of stresses and stressors."[39]

At Davis, for instance, careful attention to and recognition of beagle requirements led to constant refinement of the colony setting. Within three years, the design of the colony was such that "restraining barbed wire above the fence [was] unnecessary. Approximately 400 dogs have been raised and kept within a 5-ft. fence and none has jumped the enclosure. Animals removed from their permanent quarters appear lost and show a desire to return to their respective pens."[40] This close parallel between contemporary theoretical arguments and the historical consolidation of beagles as standard experimental animals, I suggest, foregrounds some important limitations associated with bodily, encounter-based ethics. As Elizabeth Johnson notes, more-than-human thinking has proven important in countering the political ventriloquism and hierarchical power dynamics that are often associated with advocacy, and instead emphasizing undervalued forms of agency. Due to valorizing the moment of encounter itself, however, more-than-human approaches are less useful, she suggests, in accounting for the longer histories and contexts that facilitate convivial engagements. In Johnson's terms: "The present—or the encounter—is much more than the elements found within it." Drawing on Boaventura de Sousa Santos, Johnson argues that what is needed, "rather than celebrating the richness of encounters," is instead "a 'cartographic imagination' capable of illuminating how events take shape across differential times and spaces simultaneously."[41]

Our work on the Davis colony expands on Johnson's arguments by showing how this cartographic approach needs to be coupled with an understanding of how particular histories, environments, and exclusions serve as constitutive conditions for care in the present. A focus on encounters themselves misses the role of particular histories of breeding and conditioning in enabling such encounters to occur objection-free. In the case of beagles, for instance, these histories involved the culling of disruptive animals and the ongoing molding of behavior, which (at Davis itself) continued via systematic processes of tinkering with the research space to ensure beagles were less likely to object in the future.[42] Bodily needs were certainly taken into consideration and gave rise to new forms of knowledge, but in this instance were oriented toward creating a model experimental subject: one who possessed not only the necessary physical qualities but the desired affective traits. What needs to be foregrounded, therefore, are these irreducible relations between laboratories and the characteristics of "model" or "standard" experimental animals, which recognizes the systematic shaping of their needs over time. From this perspective, the emphasis on embodied care neglects the ways that somatic relations are already shaped by particular histories that make animals easy to work with. A focus on bodily encounters, in other words, can neglect what has already been excluded from a situation in order for an encounter to take place. In contexts where certain expressions of agency have already been foreclosed, questions thus need to be asked about whether it is even possible to create space for animals to "speak back" or signify their bodily needs in ways that disrupt anthropocentric assumptions.

Though important figures for bringing concerns about the instrumentalization of care to the fore, beagles are not alone in this regard, and the issues they raise are brought into focus when turning to other sympathetic criticisms of encounter-based care, which draw attention to its limitations in troubling anthropocentrism. Van Dooren uses a detailed analysis of whooping crane conservation, for instance, to unsettle Despret's account of the work of ethologist Konrad Lorenz and pose questions about her conception of "anthropo-zoo-genesis" (or the use of bodily engagements to support communication between species). For Despret, Lorenz's work elucidates the importance of taking a careful interest in research partners by using the body as a research tool. Bodily affinities, she suggests, can be used as a means of gaining a sense of felt understanding and responsibility that is generative of new understandings.

This use of the body, for Lorenz, involved providing animals "with the most natural conditions" for them to express behavior; *natural*, in this context, is decisively not an account of a purified nature but marks a recognition of the impossibility of assuming the position of neutral researcher. Rather than treating animals as entities that can be neutrally observed, "natural conditions" in Lorenz's context refers to the process of working with the bodily attributes central to animals' modes of perception, taking these engagements seriously, and learning from them. An instance of this in Lorenz's work involved re-creating the bodily relationships between mother geese and goslings through a deliberate process of imprinting wherein "Lorenz uses his own body as a tool for knowing, as a tool for asking questions, as a means to create a relation that provides new knowledge: how does a goose become attached to its mother?"[43]

Yet while Despret suggests that the relations forged between Lorenz and his gosling entailed an openness to new ways of becoming, van Dooren points out the troubling implications of using certain forms of bodily engagement as research tools. In her analysis of Lorenz, Despret argues:

> The practice of knowing has become a practice of caring. And because he cares for his young goose, he learns what, in a world inhabited by humans and geese, may produce relations. He involves his own responsibility because he will have to fulfil the goose's needs, to be a "good mother" for it, to care for it, to walk like it, to talk like it, to answer its calls, to understand when it is scared. Lorenz and his goose, in a relation of taming, in a relation that changes both identities, have domesticated one another. Lorenz gave his birds the opportunity to behave like humans, as much as his birds gave him the opportunity to behave like a bird. They both created new articulations, which authorized them to talk (or to make the other talk) differently.[44]

Though sympathetic to these arguments, van Dooren points out that Lorenz's "technique, while good for learning, may not have been so good for geese."[45] More specifically, the process of imprinting might have created new ethical obligations and opportunities for care to emerge in human-goose relations, and Lorenz was indeed rendered capable of new thought and action by being open to the goslings' demands. What slides out of view when focusing on specific human-gosling encounters, however, is the way that future possibilities for agency have been foreclosed. Van Dooren argues, for instance, that the specific forms of bodily engagement borne out of imprinting produce a "*captive* form of life" that generates "a lifelong attachment to humans at the expense of relationships with other members of its species."[46] These encounters, in other words, take "advantage of an ontological openness to produce an altered way of life."[47]

What van Dooren draws attention to in this rereading of Lorenz, then, is that these encounters with geese enact a form of ontological exclusion. A particular way of being (the captive-goose) is materialized to the exclusion of a type of goose who is capable of having other forms of attachment to their environment or other members of their own species. Similar principles apply to laboratory beagles; just as a focus on convivial beagle-human encounters hides longer histories, what is also removed from view is the "altered way of life" constructed by these seemingly convivial encounters. What are rendered invisible are the factors that lend beagles' nonhuman charisma to laboratory research. Understanding these factors, however, is critical in understanding how the expression of this charisma has been molded in distinct ways, which affect both immediate encounters and future possibilities of relating. In the case of both beagles and cranes, what is also lost are the constitutive exclusions that lie behind convivial encounters, such as the culled beagles or "sacrificial cranes" who were used for breeding purposes. These are not merely ontological exclusions, then, but exclusions that are deeply politicized in that they materialize ways of being that are especially suited to working smoothly within technoscientific systems.

It is here, therefore, that the laboratory beagles-as-experimental-dogs trouble common understandings of what it means to stay with the trouble itself. A focus on beagles elucidates how it is not enough to acknowledge the noninnocence of certain forms of relation or the capacity of care to be instrumentalized. As laboratory beagles illustrate, ingrained histories of relation—and exclusion—can pose profound challenges for using situated, embodied relations of care to generate new forms of interest, knowledge, and responsibility. Creating space to be affected or to express agency does not take into account prior (and future) ways this agency has already been molded. The ethico-political question, therefore, is how these histories, contexts, and exclusions can be brought back into the conversation. It is in relation to the thorny set of issues surrounding care and charisma that it is useful to turn to Green Hill campaigning in further depth, in order to recuperate a notion that is often seen as fundamentally untroubling: suffering.

Charismatic Suffering

In more-than-human literatures, the emphasis on embodied care is, in part, a reaction against alternative animal liberation frameworks based on notions such as rights or suffering. While stressing its historical importance, Haraway suggests that a utilitarian, Benthamite emphasis on suffering is nonetheless not the "decisive question, the one that turns the order of things around, the one that promises an autre-mondialisation."[48] As illustrated by the Felix Campaign, Haraway's criticism of utilitarianism offers an important rejoinder in light of the way that an emphasis on animal suffering can undercut activists' aims (by leading to accusations of misdirected emotion, for instance). In *When Species Meet* it is this doubt about the capacity of suffering to unsettle anthropocentric ways of thinking and acting that underpins Haraway's push for alternative modes of ethics. Her aforementioned emphasis on situated, embodied encounters is, correspondingly, positioned as a means of fostering the sort of transformations that cannot be guaranteed by a paternalistic focus on suffering alone.

On one level campaigning surrounding Green Hill seems to bear out Haraway's wariness of suffering as an ethical framework. What is disconcertingly present in both the images from the colony and subsequent media, activist, and even legislative analyses is the distinct form of "charismatic suffering" that is utilized. Suffering can be described as charismatic in this instance due to its entanglement with the different modes of charisma identified by Lorimer, wherein factors such as the physiological qualities of particular organisms, their capacity to engage in particular corporeal relations, and their potential to engage in affective encounters give their suffering a profundity that mobilizes direct ethico-political responses. As with a conservationist focus on charismatic megafauna, moreover, an emphasis on particularly charismatic forms of suffering (such as the plight of beagles) could be tactically useful in gaining public support but runs the risk of neglecting broader ecological concerns or of perpetuating an anthropocentric logic that values organisms primarily due to their relation with the human.

Haraway's suspicion of utilitarian appeals to suffering is certainly relevant to understanding controversies surrounding Green Hill, as suffering was continuously emphasized as the central issue at stake. The trial transcripts describe how the etho-abnormalities of afflicted beagles came from eleven factors, including high temperatures and noise levels, a lack of daylight, and inappropriate bedding. Also included among these factors were particular ethological constraints that were at the center of criticisms. For instance, it was ruled that there was no chance for individual animals to "escape the external stressors also coming from their peers"; there was a "lack of area for stretching the legs . . . which would have permitted activities normal for the species"; and spaces were "devoid of the olfactory stimuli and sensory essentials for a beagle."[49] These conceptions of suffering as an ethological concern were drawn on by different groups of activists, as well as state apparatuses.

Though this emphasis on suffering might be what led to intervention in this instance, from the perspective of situated care ethics it resolves the issue far too neatly. Echoing Haraway's criticism of suffering, what was central in the condemnation of Green Hill was the application of humanist models of social justice (grounded in notions of autonomy). The facility's specific problems, from a legislative perspective, were that it inflicted suffering through impinging on "natural" behaviors that were indicative of beagle autonomy. The transcript of the trial verdict, for instance, stated that "beagle dogs are hunting dogs, with a good temperament, very bright, with a strong olfactory capacity. As a hunting dog, for physical and psychic health they must have full freedom to be able to walk and also run. . . . The expression of natural or ethological behaviour of a dog is, therefore, the result of expressive possibilities relating to mental stimulation, possibility of communication and the expression of the main sensory capacities."[50] The rights attached to beagles, then, were articulated in terms of their relations with humans, which positioned them as privileged animals worthy of special ethical concern. Implicitly humanist discourses were also present in the language used by key activist groups; the Lega Anti Vivisezione (one of the activist groups who instigated legal action against Green Hill), for instance, opened their campaign publicity by stating, "It seems incredible. Yet the dog, man's best friend, is bred and used to test drugs, chemicals, pesticides, detergents and other substances."[51] The inference, then, is that the dog's proximity to humans is what legitimizes special treatment (and by extension makes the use of beagles in research abhorrent), resonating with Lorimer's aforementioned criticism that critical-activist frameworks often only work to extend rights to "certain privileged others." Such approaches, therefore, also neglect broader shifts in the terrain of animal research, such as shifting legislative frameworks for charismatic animals that have corresponded with a decline in their use, and the sharp rise in use of far less ecologically charismatic creatures (such as zebra fish and *Drosophila*).[52]

Suffering also carries other, more profound tensions for activism. An emphasis on suffering means that from a regulatory perspective, if specific ethological needs were met, then the prosecution would not have occurred; it was the facility that was the problem rather than the use of laboratory beagles in general (indeed, this was central to the ruling against the facility). When it came to questions of killing beagles, for instance, the problem was the *way* that animals were killed and what was felt to be gratuitous levels of death; veterinary evidence foregrounded that "dogs were euthanized even though they suffered only from mild, curable diseases. . . . Some of the beagles were put down with Tanax—a drug that causes cardiorespiratory failure—without prior anesthesia, which is widely considered a less ethical way to kill them." In addition, the eleven shortcomings focused on by the trial were framed in relation to the "aggravating circumstance of causing the death of 104 beagles."[53] What this emphasis does, in Haraway's terms, is to frame the problem as one of discrete acts of killing rather than the act of categorizing certain beagles as "killable."[54]

However, wariness of suffering (or at least the conceptual implications of suffering) has not just been articulated in academic contexts. As broader debates surrounding Green Hill illustrate, navigating tensions surrounding suffering is often central to animal activism. Here, for instance, activists explicitly recognized the contradiction of anchoring a critique of animal research on the mistreatment of specific beagles, by attempting to articulate Green Hill as being representative of broader problems in the industry. UK activists involved in a long-standing campaign against another breeding site, for instance, leveraged the verdict as part of their own critical narrative against the company managing the facility—B&K Universal—which was owned by the same parent company as Green Hill. Arguments such as "B&K bosses stand trial for animal cruelty and the unlawful killing of dogs" articulated the specific issues at Green Hill as part of a broader narrative that portrayed beagle rearing as intrinsically unethical.[55]

Though appealing to narratives of suffering was tactically useful for activists in relation to the particular beagles at Green Hill, therefore, those working in other national contexts resisted too much emphasis on the specifics of the facility: the point was to contest the treatment of not just these beagles but all laboratory beagles. Hence, when Green Hill images were circulated on social media, they were rearticulated as part of broader, preexisting narratives of opposition to the use of beagles in laboratory research. Figures who played an important role in the establishment of critical animal studies as an academic field similarly situated the events at Green Hill in relation to a broader critique of the use of beagles in laboratory research, as with Steve Best's argument that "animals are just merchandise, objects to breed and sell, without the slightest scruple about pain and suffering—mental and physical—that they will suffer."[56]

To an extent, therefore, Green Hill seems to bear out Haraway's criticism of an ethics predicated on suffering, with even activists recognizing the limitations of this approach. At the same time, campaigns surrounding the facility complicate some of the existing ways that suffering has been criticized, in particular the way it has been framed as less troubling to anthropocentrism than a more-than-human emphasis on bodily encounters. The historical entanglement of human-beagle relations means that rejecting suffering in favor of situated care ethics is not a straightforward matter. In the context of laboratory animals, substituting suffering narratives for an ethics of felt responsibility can mask the exclusions that lie behind processes of anthropo-zoo-genesis. To contextualize the Green Hill beagles in line with longer histories is not to sideline the importance of care in improving conditions for specific animals, but to suggest that recognition of noninnocence is in itself insufficient in addressing how care itself can be used to foreclose the capacity for animals to impose obligations on researchers. The tensions surrounding these conceptions of care, of course, are aside from broader criticisms of animal research that have been articulated from critical-activist perspectives.[57]

Staying with Suffering

Embodied care has been seen as a means of staying with the trouble, to the point (highlighted in the previous chapter) that alternative care frameworks have been marginalized. It needs to be recognized, however, that this form of care can also be fundamentally *untroubling* and, in certain instances, can actively shore up anthropocentric frameworks or instrumental relations by undermining perspectives (such as narratives of suffering) commonly used to challenge them. The longer histories of relations between species, which have such sharp implications for care-in-the-present, can become lost within both utilitarian narratives of suffering *and* situated appeals to care. From this perspective it is unhelpful to suggest that one approach "turns things around" more radically than the other.

What I suggest needs further exploration, in light of these issues, is whether trouble can be generated not by rejecting suffering but by staying with this much-criticized framework a little longer. Suffering is not a homogeneous entity, and its role in ethical practice is more troubling than is acknowledged by blanket criticisms of utilitarian animal liberation frameworks. The distinct charisma of canine suffering, for instance, proved particularly potent in holding together concerns about the Green Hill beagles, which had been articulated by different interlocutors, in a way that gave the impression of onto-epistemic stability and created space for intervention. A closer examination of the campaigns reveals the sheer amount of *work* involved in stabilizing the matters of concern at stake at Green Hill and the critical role of suffering in achieving this stability.

Through establishing a shared object of concern, suffering beagles, a temporary alignment emerged among animal liberation actions, welfarist groups, public prosecutors, veterinarians, and legislative frameworks, an alignment that was grounded in a particular conception of beagles' ethological requirements and—crucially—in an affective engagement with their charisma as the images circulated through social media. In Annemarie Mol's terms, ethical stability was grounded in a particular form of ontological politics wherein a range of actors were brought together to enact the issue in particular ways.[58] A whole assemblage of different sociotechnical actors was mobilized to provide evidence to answer the question of whether the animals suffered; this evidence suggested the answer was "yes," and then the facility was dealt with accordingly.[59]

The stability afforded to the reality of suffering, in this instance, was thus in itself the product of choreography between the different groups involved, wherein understandings of what occurred at the facility and—crucially—how these events were to be interpreted and acted on appeared as a neat timeline. From a certain perspective, it seems as though particular realities drive particular actions: bad practice at the facility led to sustained campaigning work, which culminated in the April beagle release; images from the release were circulated on social media and garnered public and, by extension, political attention; police raided the facility as a result of this attention; expert evidence was then gathered, which, finally, led to prosecution. This is precisely the order of events that was captured on the activist-produced timeline that opened this chapter, which was perpetuated by mainstream reports in outlets ranging from *Science* to the BBC as well as within trial transcripts. Suffering played an important role in holding this narrative together and enabling different actors to assemble around it.

An overly critical interpretation of activist narratives, therefore, skirts around the work that lies behind these narratives and the importance of notions of suffering in assembling and mobilizing different actors around fraught political issues. The interventions that were enabled by these narratives were also significant; activist campaigns surrounding Green Hill were cited as the motivation for the investigation and ultimate prosecution of facility staff within the final verdict, for instance. In addition to the eventual rehoming of the beagles, the campaign drew wider public attention to laboratory beagles in general, with mainstream media reports about the trial drawing on activists as spokespeople (in a break from the sort of media rhetoric discussed in the previous chapter).

Though totalizing ethical frameworks are often criticized for failing to address specificities, in the case of Green Hill tactical appeals to charismatic suffering were precisely what drew attention to individual animal bodies in this specific facility. More, these tactics were effective in making beagle bodies matter in legislative terms as well as ethical ones, as evidenced by the painstaking details about the facility's environment and the behavior of particular animals, which were provided in trial transcripts and used as the basis for the prosecution.[60] For instance, the evidence gathered in the trial emphasized the inability for beagles to socialize "naturally," something explicitly demanded by EU legislation on animal research, which automatically classifies as "severe" any research where "social" species (including dogs and nonhuman primates) are kept in isolation.[61]

Judging this campaign as successful or unsuccessful, therefore, or evaluating it on whether it troubles or reinscribes anthropocentric relations, denies its messiness. The apparent stability of the Green Hill timeline masks fundamental differences among different stakeholders about what it means to suffer. Though, to an extent, suffering worked to consolidate a temporary ethical narrative, in practice beagle suffering was not a stable entity but (in Mol's terms) enacted in multiple ways, in being understood and performed very differently by the different groups involved. Alliances among disparate groups that helped to create beagle suffering as a stable object were temporary and contingent, with the very factors that led to this stability also lying at the center of ethical rifts—particularly the emphasis on beagle ethology. This multiplicity of understandings of suffering that were held together in order to create grounds for shared understanding and intervention, however, does not need to be read in a negative light. Instead, this diversity means that, in practice, complex and contradictory constructions of suffering such as this exceed narrow utilitarian notions and create space for more complex and unsettling ethical questions.

Further Troubles

What needs to be recognized, therefore, is that neither suffering *nor* bodily modes of care—to frame things in Haraway's terms—fully "turns things around" in itself. Indeed, in order to meaningfully depart from anthropocentrism, it seems vital that one form of care is not valorized or seen as intrinsically more troubling or radical than another. Instead, different approaches need to remain in fraught dialogue with one another in order to recognize that the contradictions inherent in each approach mark imperfect responses to an equally messy and contradictory ethico-political terrain.

For instance, although narratives of suffering should not be dismissed out of hand, it is still important to dwell on the insights offered by embodied care ethics. Both abolitionist stances toward animal rights and approaches that frame ethological arguments in terms of suffering can slip into liberal individualism. It is individual autonomy that is often positioned as important in both instances; the key differentiating factor is whether this autonomy is weighed up as a zero-sum game of "greatest good for greatest number" (as with early animal liberation frameworks such as Singer's) or grounded in values of "total liberation" wherein animals have inviolable rights.[62]

From the perspective of relational care ethics, therefore, an emphasis on suffering might be tactically useful but holds the danger of reinscribing anthropocentric hierarchies. Activist narratives that condemn the use of beagles in research on the basis of their being "man's best friend," for instance, have themselves been criticized from a more-than-human perspective, for giving the illusion that ethical relations will be guaranteed through artificially separating out certain actors as rights bearers and ignoring the messy and complex ways that human and more-than-human worlds are entangled. A focus on beagles might also neglect a host of other actors who do not end up being lucky enough to bear rights, or mask microviolences toward the privileged species by suggesting ethical needs have been met simply by avoiding more overt forms of violence.[63] Criticisms of suffering also resonate with a broader wariness of totalizing ethical frameworks that appear to offer moral closure. The key criticism is that appeals to such frameworks infer that, as long as they are followed to the letter, then this is all that is necessary; the specific needs of specific animals in specific contexts that might exceed or complicate overarching ethical standards could thus be neglected.

Yet for all the criticisms that have been leveled at critical-activist approaches, something in them remains troubling, something that cannot and, indeed, should not be dismissed out of hand. In theoretical terms, situated modes of relating that emerge through specific encounters are valuable in attuning unlike bodies to one another in order to generate novel forms of responsibility and care. In environments such as Green Hill, though, existing sociotechnical assemblages might leave little capacity for animals to impose their requirements on researchers in the way that is necessitated to support a somatic care ethics. For theorists who have advocated bodily relations as a site of care and responsibility, a lack of space for such relations is precisely the problem, hence calls across animal studies and the environmental humanities to prioritize the creation of *room* to care for individual animals.

What the history of beagle research suggests, however, is that creating space for embodied modes of care in the present does not guarantee ethical or epistemological transformation. It is not just a lack of room for care that can pose difficulties for embodied ethics, as this does not take into account the instrumental capacities of care itself. As illustrated when turning to longer histories of beagle breeding, care predicated on encounters cannot necessarily disrupt processes of "making killable" due to histories that—in certain instances—have already foreclosed expressions of agency that could disrupt predetermined experimental goals. Embodied care, in other words, might not be as troubling as it seems.

Beyond care politics itself, issues brought to light in relation to Green Hill have wide-ranging implications. Tensions between the apparent "success" of particular tactics and the sort of politics called for by nonanthropocentric perspectives speak to perhaps the most difficult fault line between theory and practice. The problem, at present, is that modes of ethics and politics that are commonly seen as the most troubling to anthropocentrism, within theoretical contexts, often directly clash with activist attempts to make particular animal bodies matter. While activist tactics do not sit easily with attempts to move beyond the human, such approaches can nonetheless be politically successful, especially in a (social) media economy oriented around capturing user attention. No matter how disquieting from the perspective of theory that has pushed for less anthropocentric modes of praxis, therefore, it remains the case that the high-profile articulation of vulnerability and suffering can serve as a powerful means of drawing public and, in this instance, judicial attention. In addition, these constructions can create a stable object for stakeholders to assemble around, in order to attribute causal responsibility and support intervention.

As Hollin and I have argued, and as elucidated by van Dooren's sympathetic critique of Despret, celebrating the ethical potential of relations and encounters not only fails to consider but can actively mask the constitutive exclusions that enable these encounters to take place. These invisible exclusions are critical to draw attention to because of the provocations they offer not just for relational care ethics but for broader questions of what staying with the trouble really means. As argued throughout the book, any course of political and ethical intervention necessarily materializes certain ways of being at the expense of others. In focusing on specific issues surrounding care ethics, what I have sought to do in this chapter is point to the broader problems that arise when the constitutive exclusions bound up with certain conceptions of trouble itself become submerged. As illustrated in previous chapters, political and ethical approaches that *seem* open are especially vulnerable to this problem; such approaches can *appear* to create space for future change, without recognizing that certain forms of contestation have already been rendered impossible. In contrast, for all of the tensions surrounding certain forms of activism, oppositional stances with clear ethical commitments at least bring their exclusions to the fore.

At the same time as I emphasize the messiness of activism, I do not want to treat appeals to nonhuman charisma in an uncritical way or indeed justify essentialist and totalizing frameworks that reinforce inequalities and offer closure. I do, however, argue that critical purchase on these approaches is impossible if they are dismissed out of hand, without careful attention to the productive and ambivalent work they can accomplish. The final chapter, therefore, pushes harder on issues that surround the politics of certain forms of representation, with the aim of recognizing their potentials while asking how these potentials can be redirected to less anthropocentric—and more intersectional—ends.

6 | Ambivalent Popularity

All worthy animals are a pack; all the rest are either pets of the bourgeoisie or state animals symbolizing some kind of divine myth. The pack, or pure-affect animals, are intensive, not extensive, molecular and exceptional, not petty and molar—sublime wolf packs in short. I don't think it needs comment that we will learn nothing about actual wolf packs in all of this.
—Donna Haraway, *When Species Meet*

Those of you who watched the *Planet Earth II* finale on Sunday night will have seen the baby turtles in trouble. The bright city lights confused the hatchlings, sending them the wrong way—towards the dangers of town, instead of the sea. But there's no need to worry, as all the crawling little turtles featured in the episode were saved!
—Response to wildlife documentary series *Planet Earth II* by children's current-affairs program *Newsround*

The depictions of suffering and dramatic image events of environmental destruction discussed in the previous chapter offer an especially contradictory form of politics. The aim of this chapter is to deepen an understanding of these sorts of contradictions and argue that certain texts can trouble anthropocentric norms not despite but because of the ambivalent politics they articulate. As illustrated by the above report from *Newsround*, representations of animals on-screen hold the capacity to elicit powerful emotional responses—in this instance, responses so powerful that program makers felt compelled to respond to the audience's anxieties about flagship nature documentary *Planet Earth II*.[1] While "little crawling turtles" seems to reflect sentimental, Disneyfied appeals to charismatic animals, the emotions generated by this imagery should not be dismissed lightly. As argued in the previous chapter, appeals to charismatic megafauna offer particular dangers for environmental politics, but they also open up possibilities for awareness raising and intervention. Dismissing these forms of representation outright not only fails to attend to their ambivalence but can make it difficult to gain meaningful critical purchase on how they *work*, which is dangerous in light of their visibility and influence.[2] It is instead necessary to pay attention to the contradictions within popular environmental texts, in order to try and recuperate the ethical and political potentials of imagery that, on the surface, seems unhelpfully sentimental or anthropomorphic. In order to take these texts seriously, however, it is necessary to situate these questions of representation in relation to longstanding traditions within media and cultural studies.

Though important ground has been gained by recent attempts to reclaim, valorize, and assert the politics of emotion, the sentimental is often still treated with derision.[3] The objects of this derision are broad-ranging: from the gendered conflation of certain emotions with irrationality, to the routine symbolic violence that has historically been leveled at working-class culture.[4] As hinted at by Donna Haraway, sentimentality toward animals in particular has been framed as a meek, petty-bourgeois counterpart to more affective, pre-emotional intensities. *When Species Meet*, for instance, begins with an overview of other attempts by contemporary theorists to engage with more-than-human worlds, and (as the opening quote illustrates) Haraway is particularly stinging about the "wolf/dog" opposition employed within *A Thousand Plateaus*, which pits the homely against the radical alterity of wolf packs.[5] While recognizing that the aim of Gilles Deleuze and Félix Guattari might not have been to grapple with more everyday concerns, Haraway nonetheless asserts that "no reading strategies can mute the scorn for the homely and the ordinary in this book," before continuing, "Little house dogs and the people who love them are the ultimate figure of abjection for D&G, especially if these people are elderly women."[6]

Debates about sentimentality and the popular thus speak to some of the recurring questions of this book; as described in previous chapters, certain communities are regularly excluded from debate due to their propensity to care in the "wrong" way. These questions of exclusion become more pronounced when it comes to cultural preferences, where accusations about the mawkish sentimentality of particular media texts leave little room for the agency of those who engage with them. Matters of cultural discrimination, moreover, are matters of taste and hence inextricable from particular constructions of class and gender. The classed and gendered dynamics of aesthetic value has, correspondingly, been central to the study of popular culture, with the tastes of particular communities often subject to totalizing critique (and these value judgments criticized in turn).

For instance, to somewhat crudely summarize over seventy years of theoretical debate: while Theodor Adorno and Max Horkheimer's staunch critique of the culture industry has been integral in establishing the political importance of popular culture, and its worth as an object of study, their work has also offered a foundation for cultural theorists to react *against*. Their most famous work, *Dialectic of Enlightenment*, is littered with memorable arguments such as "Amusement under late capitalism is the prolongation of work. It is sought after as an escape from the mechanized work process, and to recruit strength in order to be able to cope with it again. . . . The ostensible content is merely a faded foreground; what sinks in is the automatic succession of standardized operations. . . . No independent thinking must be expected from the audience: the product prescribes every reaction: not by its natural structure (which collapses under reflection), but by signals. Any logical connection calling for mental effort is painstakingly avoided."[7] Although the influence of this characterization of popular culture cannot be underestimated, it has also offered the ground against which subsequent cultural theory (especially work focused on active reading, subcultural politics, and creative consumption) has reacted. Indeed, any comprehensive text about the evolution of cultural studies or the development of audience research will offer an overview of the trajectory of these debates about the popular, in which work from the Frankfurt school still plays a critical and contested role.[8] While the above is only a thumbnail sketch of important developments within the analysis of popular culture, therefore, it nonetheless offers context from which to make sense of more recent discussions about the mediation of *environmental* politics in which a similar suspicion of sentimentality has shaped the contours of debate.

To put things in simple terms, sentimentality is often understood as being of a piece with anthropomorphism and hence as offering a somewhat shaky foundation for an ethics that seeks to go beyond the human. Criticisms of sentimental representations of animals have crystallized these arguments, because such depictions are seen to perpetuate the sense that animals are worthy of attention as long as they are somehow "like" humans.[9] In drawing on Deleuzian affective logics to inform his own analysis of film, for instance, Jamie Lorimer sees an affective logic of sympathy as offering an especially problematic mode of representation, due to relying on "gaudy" anthropomorphic imagery.[10] Furthermore, as discussed in depth below, he is not by any means alone in making these arguments. Indeed, metonymic slippage between sentimentality, anthropomorphism, and anthropocentrism underpins criticism about particular representations of animals on-screen. It is precisely this set of associations, however, that I suggest demands closer attention, especially in light of longer histories where matters of aesthetic judgment and categorization are bound up with hierarchy and exclusion.[11]

Attending to debates about popular cultural representations of animals is particularly urgent in light of a contemporary political moment that has seen an upsurge in what Mike Goodman and colleagues describe as "spectacular environmentalisms."[12] The notion of a spectacular environmentalism is evocative and captures something important related to the problems focused on within this chapter. The Situationist connotations of the term *spectacle* point to a conception of the mainstream media as offering distraction from material inequalities, a distraction that can be combated only through active cultural resistance, such as the production of subvertising and alternative media.[13] In referring to the mediation of environmental politics as spectacular, Goodman and colleagues thus point to ongoing concerns about the commodification and depoliticization of ecological issues.[14] The celebrity promotion of environmental causes or representations of ecological catastrophe in nature documentaries, for instance, could be seen as transforming urgent issues into entertainment, with depoliticizing consequences.[15] These representations might prompt sensations of pity, concern, or even horror but (arguably) rarely translate into action. More, the mediation of environmental politics actively reinforces the systems and social relations that lie at the root of these problems; the critical purchase of popular environmental media, for instance, is often constrained by advertising revenue and the interests of media company owners, among other factors.[16]

Yet though Goodman and colleagues' use of the term *spectacular* does point to issues of commodification, their framework also takes care to emphasize that ideology critique in itself is insufficient in grasping the complex and contradictory political work accomplished by popular culture. Rather than reading spectacular environmentalisms as straightforwardly apolitical or self-defeating, therefore, they instead emphasize their *ambivalence* and the ways "these politicized media processes influence a range of equally politicized ways of seeing, being with and relating to diverse environments through a tethering of the spectacular to the discourses and practices of the everyday."[17] Spectacular environmentalisms, from this perspective, are never straightforward and can simultaneously offer ethical provocations while reinforcing norms that are grounded in liberal-individualist values. In focusing on media that belong to this broad genre, therefore, this chapter draws together debates that have recurred throughout this book but approaches them through a slightly different lens.

The Politics of the Popular

Within this book I have progressed broadly chronologically, from anticapitalist activism in 1980s campaigns against McDonald's to present-day circulations of emotive imagery via social media. At the same time, each chapter of the book has retained a focus on recurring problems facing nonanthropocentric praxis and some of the messy ways that key tensions have been negotiated in grassroots political settings. Throughout I have drawn attention to the difficulty of highlighting who or what is excluded from particular sociotechnical arrangements, while also finding ways of being responsible to the exclusions that arise in alternative infrastructures and practices. Tensions faced by the activists I have discussed, moreover, are not simply practical matters but offer provocations for theories that have emphasized more-than-human entanglements and relational ethics.

These provocations are intensified in this chapter, which works against the backdrop of debates about *popular* culture to explore the difficulties that arise when particular values become popularized.[18] It might seem paradoxical, or even churlish, to frame tentative successes as problematic. As I go on to discuss, recent documentaries such as *Cowspiracy* seem to mark a cultural moment where issues that were formerly the marginal concerns of fast-food activists have gained mainstream visibility. At the same time, the issues at the heart of this chapter elucidate that often it is precisely the moments when ideas, values, and practices gain broader currency that pose the thorniest questions.

In focusing on problems of popularity, this closing section of the book does two things. First, and most simply, the chapter brings things up-to-date by providing an overview of more contemporary developments within popular media culture, where a number of existing themes in the book converge and where further uneasy affinities between relational theories and activist perspectives can be located. The second, and more important, aim is to negotiate the distinct set of problems that can arise when activist practices gain a degree of traction within popular culture. The context for these debates is a cultural environment in which the tactics engaged in by anticapitalist activists resonate with popular media that range from prime-time wildlife documentaries to celebrity-fronted awareness-raising films, and in which dietary practices that, likewise, formerly signified countercultural resistance (in the context of protest camps and food activism) have become a prominent feature of urban bohemia. Defining the difficulties outlined within this chapter as a problem of popularity is, moreover, a deliberate gesture that situates them in relation to debates about popular culture within cultural and media studies that have long foregrounded the ambivalence of the popular.

In making these arguments, the chapter thus draws together different strands of feminist theory; while revisiting concerns from feminist science studies that have been explored throughout the book, it also engages with a slightly different set of debates in feminist media and cultural studies that have explored the relationship between feminism and popular culture. Engaging with the latter body of work is important. While feminist media and cultural scholarship has always drawn attention to the ways that popular culture can reinscribe social and cultural norms, it has also complicated simplistic assertions about the ideological values and discourses found within particular texts. Recognition of the messiness and ambivalence of popular culture is especially useful in nuancing arguments that have been made about the mediation of environmental politics within nonanthropocentric theoretical contexts.

Rosalind Gill's account of postfeminism as a *sensibility* offers an especially helpful way of thinking through contemporary media representations of anthropogenic problems. Gill uses the notion of sensibility in reference to postfeminist media culture in order to avoid framings of postfeminism purely as a "backlash" or a rigid "epistemological perspective."[19] It is necessary, Gill argues, to avoid either using *postfeminism* as a label to designate media that are a simple backlash against feminism or treating it as a coherent aesthetic that is expressive of hegemonic femininity. Reality television programs focused on makeovers, pampering, and "girls going wild," for instance, might give the impression of a trend within popular media where feminist narratives of collective change have been displaced by notions of individual empowerment, yet such a reading obfuscates the less straightforward ways these texts operate.[20] Gill suggests instead that postfeminism should be conceived as a sensibility, a mood established by a series of recurring themes and elements within popular culture that orient its ambivalent politics. The elements that constitute a postfeminist sensibility might superficially give the appearance of coherence, but a closer examination reveals productive tensions (where narratives of empowerment and individual choice sit uneasily with the dominance of makeovers as a paradigm, and where an insistence on women's sexual agency is coupled with the resurgence of ironic sexism, for instance).[21] Gill suggests, therefore, that these texts consist of an "entanglement of feminist and anti-feminist themes within media texts."[22]

Prominent cultural theorists such as Angela McRobbie have similarly drawn attention to these entanglements of postfeminism in a critical way, characterized by a tendency to evoke feminist themes only to undermine them.[23] Postfeminism, in other words, either commodifies feminist values or depoliticizes them (by focusing purely on individual empowerment, expressed as consumer choice, rather than on collective change).[24] More recently, however, a body of work has emerged that has read postfeminism's entanglements in a more productive way; though not uncritical, a number of scholars have suggested that the entanglement of feminist and antifeminist impulses generates tensions that are significant in that they lay bare the difficulties of articulating feminist politics within the constraints of a commercial media environment. Wallis Seaton, for instance, draws attention to the intensely compromised nature of the feminist politics that exists in popular culture, in ways that resonate with Goodman and colleagues' analysis of spectacular environmentalisms. The high-profile texts Seaton focuses on (from the *Hunger Games* trilogy to Lena Dunham's *Girls*) certainly display the type of entanglements described by Gill and foreground the danger of reducing feminism to a brand or selling point. Yet Seaton argues that the contradictions contained within these texts are nonetheless generative in being the site of fraught debate, which can make particular gender inequalities—notably those related to the emotional labor of young women—visible and open to contestation.[25]

There are several reasons why it is productive to read these understandings of the entangled nature of the postfeminist media landscape against relational theories that have emphasized more-than-human entanglement and interdependency. First, to echo my argument in previous chapters, it is important to move beyond identifying particular practices, representations, or forms of politics as either definitively anthropocentric or nonanthropocentric. As noted throughout the book, such judgments can ignore some of the material constraints on practices of articulation that lead to compromise, as well as the hard work and creativity involved in navigating these constraints in practice. In the context of popular media, moreover, clear-cut judgments about whether something perpetuates anthropocentric logic can mask the ambivalence of imagery that—on the face of it—seems undisputedly anthropocentric or sentimental. It is also important, second and relatedly, to be wary of making implicit value judgments based on the emotional resonances that are associated with particular texts; as noted above, and echoing the previous two chapters, the sentimental is often (implicitly or explicitly) denigrated, and this is especially true in analyses of popular texts.[26] Feminist media and cultural studies offer important resources for recuperating the sentimental and refusing to exclude its politics out of hand. Third, and finally, the use of the term *entanglement* by feminist media and cultural scholars is especially productive and requires further engagement.

The lineage of entanglement in work such as Gill's is slightly different from the genealogy drawn on in previous chapters in relation to Karen Barad. If arguments made about postfeminist media are read through a Baradian lens, however, then these debates are even more valuable for grasping how the ambivalent qualities of spectacular environmentalisms operate. If uses of entanglement in feminist media studies are taken not to mean the straightforward tangling of—or messy relations between—competing tendencies (that could potentially be disentangled), but instead as pointing to the mutually constitutive relationships *between* these contradictions (in a Baradian sense), then this offers a helpful framework for interrogating environmental politics within popular culture. A focus on these entanglements, in other words, necessitates a deeper understanding of how opposing anthropocentric and non-anthropocentric tendencies work through and coconstitute one another. In reading these bodies of work against one another, I draw attention to not only the exclusions that are fostered through textual entanglements but (building on the previous chapter) the invisible exclusions that are fostered by theoretical *analyses* of these texts.

As a means of foregrounding the productive ambivalence of the popular, throughout this chapter I combine the insight that has been offered by feminist media and cultural theorists with an approach more familiar to feminist science studies, the use of figurations. Figures are central to Haraway's approach, from her famous cyborg and Oncomouse™, to figuring ethical engagements through a focus on her dog, Cayenne, in *When Species Meet* and her more recent figure of Chuthulu. Haraway sees figures as valuable due to their capacity to offer "material-semiotic nodes or knots in which diverse bodies and meanings coshape one another."[27] As I have argued with colleagues in our own exploration of the politics of figurations, the very *being* of figures renders normative distinctions between animate and inanimate, nature and culture, or animal and human untenable; for this reason, figures have proven especially valuable within the environmental humanities, where they have offered an important means—as Michelle Bastian puts it—of intervening "into habitual ways of both living in and understanding the world in order to denaturalize the commonsense feel of conventions and open them up so that things may work differently."[28] The remainder of this chapter takes a lead from this approach to using figurations, focusing on three in particular—elephant, turtle, and cow—who offer especially helpful nodes for drawing together affective concerns, popular representations, and theoretical debates about the sort of environmental politics that can emerge within the productive constraints of popular culture.

Elephant: Sentimentality, Anthropomorphism, and Cultural Hierarchies

Elephants offer helpful coordinates for debates about spectacular environmentalisms and epitomize heated discussions that have emerged in relation to the mediation of more-than-human agency. Particular elephants, and elephant communities, have acted as flash points for debates about sentimental representations of animals, as well as the links among these representations, anthropomorphism, and anthropocentrism. Elephants are thus particularly helpful figures for understanding tensions that have arisen in discussions of animals on-screen, and are also valuable in situating these debates within specific media histories.

As especially charismatic animals, elephants have been represented in a diverse range of ways; Lorimer, for instance, draws on different representations of elephants to delineate four affective logics, which he argues structure audiences' affective responses to mediated representations of animals. For Lorimer certain affective logics reduce more-than-human agency to an anthropocentric model, where animal desires are made intelligible through anthropocentric tropes. He sees Disney's *Dumbo*, for instance, as typifying a "sentimental" affective logic in its universal themes, evocative animation (which emphasizes relatable elephant features, such as "eyes, face, and hands"), and stylization that "works with a gaudy logic of sensation."[29] This anthropomorphism, by extension, reproduces a reflective logic wherein the motif of the elephant is simply a projection of decidedly human subjectivity: "Dumbo as allegory [thus] reduces the alterity of emotional, living elephants to an anthropoidentity."[30]

Other, more productive affective logics—of sympathy and awe—can be found in specific instances of activism and wildlife documentaries focused on elephants, although these too have their limitations. Lorimer argues that documentaries adopting a biographical approach, which foreground the plight of specific animals, share a similar logic of sympathy to awareness-raising films produced by charities such as PETA. Both types of text, he suggests, "are didactic and moralistic and have a strong political message" that is often reinforced by emotively charged imagery.[31] Lorimer is certainly not alone in raising concerns about these forms of representation, with his arguments again echoing a number of thinkers—in science communication as well as media and cultural theory—who have argued that despite soliciting an initial "shock to thought," an affective logic of sympathy is ultimately constraining, as shocking imagery can quite simply be exhausting and result in apathy rather than action.[32] Nature documentaries that operate on a different affective register of awe have similar constraints. An awe-generating approach is commonplace in documentaries that seek to present themselves as a neutral window on the natural world—and use ever more advanced technologies to do so.[33] Lorimer's sympathetic criticisms of this approach reiterate the sort of concerns raised within feminist science studies about the danger of presuming a bird's-eye position of neutrality. While this approach evokes awe of the "overwhelming size, power and alterity of nature," the human is inevitably positioned as an impartial observer (or at best admirer) of this world.[34]

In contrast with sentimentality, sympathy, and awe, the most productive logic, Lorimer argues, is an experimental one that evokes curiosity through conveying the alterity of animals. What is critically important in terms of whether particular representations reinforce or undercut anthropocentric norms, he suggests, is the capacity of representations to be affective without resorting to sentimental anthropomorphism. This approach, however, is almost always confined to avant-garde cinema, as reflected by his argument that "the micropolitics of disconcertion expressed in experimental media operate in different registers to the cloying sentimentality, sympathetic outrage, or awe-ful respect of the previous three genres."[35]

This burgeoning focus on affective logics within film, however, needs to be approached with caution. The argument that different affective logics are connected to different degrees anthropocentrism can reinscribe worrying cultural hierarchies. The categorization of texts according to their affective logics creates an implicit hierarchy about *which* forms of representation and constructions of animal subjectivity are affectively (and by extension ethically and politically) productive and which are problematic in flattening more-than-human agency and maintaining human exceptionalism. The cultural politics of this hierarchy is concerning in light of aforementioned debates about popular culture, echoing the "scorn for the homely and ordinary" that Haraway identifies in Deleuze and Guattari. Furthermore, hierarchy is not the only issue with a focus on affective logics, and questions need to be asked about whether this emphasis can fully address the productive ambivalence of popular culture. While a focus on affective logics can be helpful in examining how particular texts are structured, as a number of critics of affect theory have argued, care must be taken when assuming that these logics correlate directly to specific emotional responses, let alone actions.[36] A messier approach is thus needed that draws inspiration from feminist media and cultural studies and addresses some of the dangers posed by affective hierarchies, as well as recognizing the coexistence of (and clashes between) particular logics.

To illustrate these arguments, the rest of this section turns to an especially prominent figure who is difficult to reduce to any one reading: Topsy the Coney Island elephant. The public electrocution of Topsy in 1903 has served as a touchstone for a growing body of research about mediated environmental politics, due to being documented in the short (seventy-four-second) film *Electrocuting an Elephant*.[37] After killing a spectator who had deliberately burned her trunk with a cigar, Topsy was constructed as a rogue elephant and condemned to execution by hanging, poison, and electrocution.[38] The filming of Topsy's death has since rendered her an important figure for materializing the entwinement of commercial interests, entertainment, and animal capital within the emerging cultural industries.[39]

Despite its brevity *Electrocuting an Elephant* is a multilayered text that has been read in a number of ways: Rosemary-Claire Collard has argued that Topsy's death crystallizes the noninnocence of wildlife on-screen, with the violence of this film encapsulating subsequent coercive engagements with animals in the cultural industries.[40] The execution has also been seen as a model for the carceral and corporal punishment of humans; indeed, there has been heated debate about *which* model it offers.[41] Anat Pick has similarly drawn attention to violence, but with a slightly different emphasis in foregrounding how Topsy's spectacular death makes vulnerabilities shared between human and animal bodies visible in ways that open ethical obligations.[42] Nicole Shukin's nuanced analysis furthers these lines of argument by exploring the different layers of the film that become apparent when its experimental dimensions are focused on.

Shukin argues, for instance, that the experiment had a dual purpose, simultaneously illustrating the mortal danger of direct current (thus suggesting that the system promoted by Thomas Edison's rival George Westinghouse was unsuitable for domestic use) while establishing electricity in general as a tool for quick, painless execution (that was so reliable it could instantly fell an elephant). In taking this focus, Shukin draws attention to the instrumental ways that animal affect was utilized within the film as a means of communicating Edison's arguments, suggesting that "while animals have been barred from the logos and the domain of the symbolic in discourses of Western modernity (from 'telling') they have nevertheless been conceived as eloquent in their mute acts of physical signing and their sympathetic powers of affect (in 'showing')."[43] Yet the process of rendering Topsy a transparent sign was not by any means straightforward. Topsy's well-documented acts of resistance toward trainers challenged her construction as a circus performer, and the filming of *Electrocuting an Elephant* was shaped by her refusal to proceed toward the execution apparatus (with filming forced to stop for the two hours it took to move the wiring and scaffold to Topsy rather than vice versa).[44] These expressions of resistance, however, were ultimately either used to construct her as dangerous and to legitimate her death or rendered invisible through being edited out of the film itself.

Thus, while an analysis of the broader sociotechnical relations bound up with Topsy's death draws attention to, as Maan Barua puts it, the *constitutive* role of animal bodies within regimes of lively capital, it is also important to consider the *constraints* imposed on this agency that might not immediately be visible.[45] To reiterate Elizabeth Johnson's arguments about the importance of grasping the broader relations and histories that constitute particular encounters, it is necessary to pay careful attention to the relationship between the film's affective charge and the broader histories that framed *Electrocuting an Elephant*. In this instance it is necessary to attend to Topsy's imbrication in the legacies of colonialism (which allowed her to be transported to the United States as a circus animal), urban arrangements (in the form of the shifting leisure pursuits that saw the emergence of Coney Island), the development of modern infrastructures (as with the role of electricity), and popular culture (i.e., the need to pay attention to the distinct relations of the culture industry during this period).[46]

The multitude of different readings offered by Topsy's electrocution are, therefore, I suggest, due to the inherent tensions contained within the film itself, which are bound up with *Electrocuting an Elephant*'s sociohistorical and technical environment. To go back to questions of entanglement: in order to make sense of the film, it is necessary to understand the way that particular representations of animals on-screen are not simply a grab bag of contradictions; rather, the opposing tendencies in such films work through and actively co-constitute one another. For instance, the affinities constructed between Topsy and certain humans (criminals, those worthy of punishment) are what allowed her to be "animalized" and rendered killable (and indeed vice versa). In some instances, to echo Gill's and McRobbie's accounts of postfeminism, the different logics in the film also undercut one another, as when Topsy's disruptive expressions of agency—and the impossibility of controlling her—legitimized her death, or when shared vulnerability between species was realized through the demonstration of human mastery over nature.

Even a text such as *Electrocuting an Elephant*, a film less than two minutes long, thus foregrounds the danger of categorizing representations—and the emotions or ethical resonances that they generate—in straightforward ways, as this can smooth over informative and co-constitutive tensions. In more recent texts, especially those that advocate more explicit forms of ethical action, moreover, it is the entanglements between seemingly contradictory ethical impulses that are especially important. It is therefore critical to understand how the specific way that these entanglements are materialized can create certain ethical potentials while undercutting others. The rest of this chapter, correspondingly, works to conceptualize how anthropocentric and non-anthropocentric aspects of texts cut against each other in ways that afford sentimental, awesome, and sympathetic imagery a more complex politics. To develop these arguments, I turn to two particular figures who are especially useful in plotting a path through the contradictions of spectacular environmentalisms: turtles and cows.

Turtle: Ethical Obligations and Interventions

Isabelle Stengers's use of the phrase "turtles all the way down" is a refrain that recurs throughout Haraway's work and has been picked up by a number of other thinkers in order to point toward the irreducible complexity of the world.[47] *Specific* turtles have also played an important role in conceptual debates; particular turtles and turtle species have been attended to within some rich theoretical work, which has gone beyond using them as a means of gesturing toward complexity, in order to ask questions regarding how to act and intervene once complex interdependencies between species have been acknowledged. Turtles, for instance, have proven important in the context of extinction studies, as in Bastian's work, where they have signaled the impact of anthropogenic problems, the need to respond to these problems, and the difficulty of doing so in a context where the disparate needs of an irreducibly complex array of actors are brought together. What happens, for instance, when habitat destruction forces species such as jaguars and leatherback turtles together, and the former acquire a taste for turtle? When both species exist at what Thom van Dooren describes as the drawn-out "dull edge of extinction," what sort of intervention is needed or even possible?[48]

Turtles as figures, therefore, have spoken to increasingly pressing tensions between more-than-human theories and environmental politics, in response to the sort of problems described by Alexis Shotwell as the difficulty "of resisting human exceptionalism while at the same time thinking that humans have responsibilities."[49] Maria Puig de la Bellacasa sharpens the stakes of this argument still further, suggesting that the difficulty of moving beyond anthropocentrism could actively be *caused* by the very forms of relational ethics that are designed to overcome it. Puig de la Bellacasa wonders, for instance, whether "the symmetrical redistribution of affective agency in complex relationalities of humans and nonhumans" could reinforce a "persistent reluctance . . . to consider (our) intervention and involvement, and let's say ethico-political commitment and obligations."[50] Puig de la Bellacasa's and Shotwell's arguments, then, point to the central issue of whether the decentering of human agency and recognition of entanglement can have the consequence not of fostering ethical responsibility (as theories of companion species intend) but of permanently delaying it for fear of reinstating humans as privileged agents. Again, these debates point to the persistent question of whether it is possible to move beyond recognition of entanglement and interdependency, in order to make some sort of critical intervention. Turtles are especially potent figures for pushing these concerns to the foreground.

From the 1950s, naturalist Sir David Attenborough has fronted a series of highly successful (as well as relatively high-budget) nature documentaries, including *Life on Earth*, *Blue Planet*, and *Planet Earth*, all of which were originally produced by the BBC but globally franchised.[51] As a subgenre of nature documentaries, these series make use of sophisticated technologies necessary to gain insight into formerly hidden aspects of animal life, to spectacular effect. As discussed above, Lorimer describes documentaries in this mold as conveying an affective logic of awe, wherein "the moving animals evoked in this register are fundamentally wild and different."[52] This characterization seems borne out by Attenborough's own account of the format: "If you're a film cameraman you are trained, as it were, to be the observer, a non participant. That's very important."[53] In 2016, however, this framing of the spectacular wildlife documentary as neutrally depicting a predefined natural world "out there" was unsettled by the journey of a small hatchling turtle in Barbados.

The turtle appeared in a sequence from the final episode of the wildlife documentary *Planet Earth II*, which was commercially successful but did not go without criticism.[54] Segueing with Lorimer's broader criticisms of the logic of awe, other documentary makers argued that the program's emphasis on the spectacular masked anthropogenic problems: "These programmes are still made as if this worldwide mass extinction is simply not happening. The producers continue to go to the rapidly shrinking parks and reserves to make their films—creating a beautiful, beguiling fantasy world, a utopia where tigers still roam free and untroubled, where the natural world exists as if man had never been."[55] The worlds evoked in these documentaries, in other words, are accused of having the "Eden under glass" air that Haraway was so wary of in the context of rain forest conservation (see chapter 1). The final episode of the series, however, seemed to hold promise for posing more difficult questions about human obligations related to anthropogenic problems. Entitled "Cities," the sixth episode of *Planet Earth II* focused more decisively on entanglements between species, along with the dangers and ethical dilemmas they pose: from the expansion of cities leading leopards to hunt in urban areas, to macaque monkeys opportunistically ransacking apartments for food. Yet "Cities" was ambivalent. On the one hand, the urban environment offered opportunities for unpicking human exceptionalism, but, on the other hand, many of these sequences contained an implicit critique of transspecies entanglements by positioning human activity as encroaching on nature. The most infamous scene of the episode, for instance, depicted baby turtles in Barbados, making a perilous journey to the ocean. The most controversial moment portrayed turtles disoriented by the lights of the city and walking in the incorrect direction—away from the shore—to die in storm drains and under car wheels.

In order to make sense of this scene, however, it is necessary to provide a little more conceptual context regarding turtles' significance. In Bastian's work leatherback turtles have proven especially productive figures for mapping clashes in temporalities, where anthropocentric rhythms cut across and disrupt those of other species. Conventional clocks, she suggests, might offer markers for human rhythms, routines, and working hours but are less useful for "indicating the wide varieties of clashing time scales and modes that characterize the present and which we need to negotiate in our responses to climate change, resource depletion, and mass extinctions."[56] Bastian's turtles help to elucidate the stakes of clashing temporalities; she asks, for instance, whether turtles could be conceived of as an alternative clock, one that measures time on a different scale than anthropocentric time. The evolutionary history of leatherbacks—in stretching back over 100 million years—offers a connection to a world that existed long before human civilization, but these turtles also illustrate how more recent histories have disrupted this continuity: "Turtles not only tell us about the unstable time of an active Earth, they also tell the frustratingly slow time of human efforts to respond to recognized environmental threats."[57] In tracing how humans have disrupted these long rhythms, therefore, Bastian is able to foreground difficult questions about human obligations. This ethical potential is illustrated by van Dooren's engagement with Bastian, when he draws on her work to foreground that what is at stake in extinction is the intergenerational labor of a species, labor that has been established over millennia. It is not, therefore, just individual animals but whole ways of life that are undone by human activities.

These debates are important to bear in mind when considering critical responses to *Planet Earth II*, particularly in episodes such as "Cities." Amid public uproar across social media about the plight of hatchling turtles, the filmmakers were criticized for apparently leaving the hatchlings to die. This led to a statement being released by the BBC that claimed, "Every turtle that was seen or filmed by the *Planet Earth II* crew was collected and put back into the sea," the reason for this being that "in this instance, the problem was man-made and it was therefore appropriate for man to step in to assist."[58] The BBC's statement thus firmly reinstates a sense of human encroachment on nature, a situation that demanded intervention from European camerapersons. A focus on turtles that were "seen and filmed" also freezes the moment in time, as a discrete moment that can be resolved by individual (human) actions.

Yet although the above aspects of the filmmakers' statement seem to reinforce the brand of paternalistic savior logic that Haraway criticizes in "The Promises of Monsters," such a reading does not wholly capture what is happening in this instance. For instance, an important dimension of audience responses to the episode was that human responsibility should not begin and end with the fact that these particular turtles were saved, with members of the public asking questions such as "What about thereafter? Any local groups monitoring?"[59] Although these concerns were raised by a minority of audience members via social media, they went on to become a dominant frame for how ethical issues raised by the series were understood, in both national and international news outlets. In response to these concerns, filmmakers made it clear that they were acting in conjunction with local activists; as well as emphasizing the ongoing work of groups such as the Barbados Sea Turtle Project in rescuing disoriented turtles—work that both preceded and continued after *Planet Earth II*—the BBC even released a new short film that documented the actions of the project.[60] What was showcased in controversies surrounding "Cities," therefore, was an environmental initiative that, in the project's own words, sought to bring together local residents and academics in order to "restore local marine turtle populations to levels at which they can fulfil their ecological roles while still providing opportunities for sustainable use by the people of Barbados, and to support similar efforts in other countries of the Caribbean."[61] In producing these online supplementary films and statements, the *Planet Earth* filmmakers thus insisted on recognizing the importance of ongoing work and engagement at a local level, in ways that complicated narratives of "conservation from above" and opened space for more searching questions about how to act and intervene in contexts where mutual interdependencies are beginning to unravel.

Like *Electrocuting an Elephant*, therefore, turtles—from Bastian's leatherbacks to the hawksbill hatchlings depicted in *Planet Earth II*—resist easy theoretical readings or categorizations. These turtles, in varied and provocative ways, are figures who show how human and animal lives are knotted together in a manner that can have dire consequences, not just for individual animals but for past and future generations. Turtles also, however, foreground the necessity of intervention, and although interventions were sometimes narrated in problematic terms that posed humans as removed from "nature" or invoked hierarchical models of advocacy, even in the case of a text as popular as *Planet Earth II* this was not uniformly the case. The picture ultimately pieced together, upon attending to the broader media ecology through which this story unfolded, elucidated the messiness of ethical engagement and ultimately foregrounded the need for situated local responses. Again, this short sequence that depicted the life cycle of a small group of turtles shows how deceptively simple texts can contain a number of contradictory tendencies. *Planet Earth II*'s turtles, in other words, foreground the importance of focusing on the mutually constitutive relationships between the contradictions that routinely exist within popular representations of animals, in order to understand how anthropocentric and nonanthropocentric tendencies can work through one another. Here the neutral bird's-eye view evoked by Attenborough, for instance, came into conflict with the episode's depiction of clashing temporalities between species, in ways that simultaneously challenged and reinforced human exceptionalism. Indeed, the clashes between the different ethical dimensions of the episode are what make it so provocative in foregrounding what is at stake in competing modes of intervention and responsibility. Even though these tendencies clashed, they nonetheless worked to open space for questions about accountability. However, while turtles might be complex, the next figure is messier still and gets to the heart of tensions surrounding the mediation of environmental politics.

Cow: The Politics of "Making Visible"

The final figures I focus on in this chapter are a series of cows, each of whom pushes questions of obligation still further, due to calling for ethical responses in a more direct and explicit way than documentaries such as *Planet Earth II*. Cows speak, in particular, to debates that have emerged about the value of activist tactics that focus on "unmasking" aspects of animal agriculture. These debates about visibility are critically important in laying bare broader tensions between nonanthropocentric theories and critical-activist approaches.

Concern with visibility has been central to work within critical animal studies that has analyzed a range of popular media and instances of material culture—from food advertising to children's toys—to explore how it naturalizes or masks particular aspects of animal agriculture.[62] Carol J. Adams's notion of the absent referent in *The Sexual Politics of Meat* (originally published in 1990) has acted as a touchstone for this line of argument, due to her assertion that the promotion of animal products relies on a detachment of material animal bodies from these products or that—when animals do appear—there is a reliance on cartoon-like depictions of the farmyard featuring animals who are complicit in their consumption.[63] Indeed, the particular marketing trend Adams focuses on is the use of sexualized images of animals who provocatively invite consumers to eat them and (to frame things in Haraway's terms) are as removed from actual animals as Deleuze and Guattari's "pure-affect animals" are from actual wolves. From this perspective, what is needed is to debunk these images and raise awareness about the material relations that underpin animal agriculture, in a move that brings animal bodies—and animal suffering—to the fore. A politics of visibility focused on suffering has also been engaged in by activists; hidden filming has been a long-standing tactic within grassroots campaigns and has resulted in a degree of success (as well as controversy).[64] Significant public debate, for example, has been generated by the release of activist films and has led to legislative change in the context of agriculture, laboratory work, and entertainment.[65] Indeed, as Claire Rasmussen notes, concern about the economic impact of filming inside farms has resulted in its criminalization.[66]

This emphasis on the political potentials of "making visible," which has been important from a critical-activist perspective, has, however, been confronted with three particular problems. First, there has been increasing concern about the sort of politics fostered by an emphasis on visibility, which is often oriented toward changing individual consumer behavior. This relates to a second, interrelated, problem: even if certain relations are made visible by activists, this does not automatically lead to political change. As Rasmussen goes on to argue, there is often undue "optimism about the ability of vision to transform ethical sensibilities." What she is pointing to in making this argument is the assumption "that only ignorance prevents a more moral position towards animals that takes account of the toll that our consumer practices take on the bodies of animals."[67] For Rasmussen, more attention is needed to the *ways* of representing animals that are offered by particular texts, but these can only be grasped by paying careful attention to these texts' contexts of production and consumption. This point is especially pertinent in relation to a final issue: activists do not operate in a semiotic vacuum but have to work within and against a shifting terrain of popular culture.

As outlined previously, the landscapes activists work within are often sutured by material-discursive relations that foreclose particular practices of articulation, while necessitating others. These relationships are complicated further by the affective logics discussed here. The past decade, for instance, has seen a shift in marketing strategies, within a broader context of concern about ethical consumption (on the part of consumers) and allegations of greenwashing (on the part of particular corporations).[68] In relation to meat consumption, these shifting marketing approaches are especially evident, as campaigns have often moved away from making animal bodies invisible—the assumption inherent in Adams's reading—and instead showcase certain aspects of animals' working lives. As outlined in previous chapters (see chapters 1 and 2 in particular), shifting strategies on the part of corporations can reframe the terrain of debate in ways that force activists to adapt their narratives, and can impose constraints that make it difficult to meet the demands of nonanthropocentric modes of articulation.

The above three problems identified by Rasmussen are not exhaustive but do illustrate that the act of making human-animal relations visible might not be enough in itself to foster the sort of material change desired by activists. Although in certain contexts, as touched on in the previous chapter, drawing attention to suffering can be a potent means of holding an issue together, such tactics need to be situated within specific media environments. To interrogate the (often highly contested) politics offered by activist imagery of animal agriculture, therefore, it is again necessary to maintain a focus on how the contradictory tendencies within these images work through one another, as this can offer better insight into the ethical significance of what is foregrounded by these texts. As with the case of leatherback and hawksbill turtles, questions about visibility and ethics are inevitably messy, and there is a risk of homogenizing some of the disparate ways that different texts have sought to depict particular aspects of agriculture (or indeed downplay specific farming practices). Here I navigate these tensions by focusing on three cow figures who are especially helpful: the titular cows of the feature-length documentary *Cowspiracy*, the cow that participants are asked to empathize with in the charity Animal Equality's virtual reality project iAnimal, and, finally, as a point of contrast, the figure of the cow-as-worker who regularly makes an appearance in food marketing campaigns.

Debates surrounding *Cowspiracy* help to encapsulate tensions surrounding the sort of politics offered by a politics of making visible. The film was initially released in 2014 as a crowdfunded documentary, before being re-released as a celebrity-fronted film on the streaming service Netflix, and both *Cowspiracy*'s content and its ready availability have been seen as an important juncture in the mainstream recognition of issues that were formerly regarded as marginal activist concerns.[69] Activist-scholar Alex Lockwood, for instance, sees *Cowspiracy* as offering important ethical provocations; like Lorimer, Lockwood draws on affect theory but takes an intensely personal approach and reaches very different conclusions. He describes his own tears at engaging with *Cowspiracy* itself and points to the way that curiosity is not purely generated by films that point to the alterity of animals (as with the logics Lorimer identifies) but could also be provoked by more surprising representations.[70] *Cowspiracy*, Lockwood argues, is particularly generative in terms of the affective engagements it fosters from a combination of shocking and sentimental imagery. The film's plot involves a protagonist (director and narrator Kip Anderson) striving to be more sustainable in his everyday life. As the documentary progresses, Anderson becomes aware that all of the actions he undertakes as personal engagement with climate change (being more energy efficient, avoiding flights) pale in comparison to reducing meat consumption. Anderson's own lack of awareness about the environmental impact of meat consumption and about the role of agribusiness in maintaining this ignorance, plus the sense that large nongovernmental organizations are to an extent complicit (due to focusing on less controversial targets), is what lends the film its name.

Describing a screening of *Cowspiracy*, Lockwood identifies two moments when the audience responded both collectively and affectively. In the film Anderson initially explores alternatives to large-scale animal agriculture by visiting a smallholding that seems to offer a more sustainable model of farming, where he witnesses the slaughter of a duck. The graphic depiction of this scene, for Lockwood's audience at least, meant "many of the audience members turned their faces away, or cried out."[71] The following scene, however, evoked a very different response. After visiting a second smallholding, instead of watching another slaughter, Anderson leaves with a chicken, then delivers her to an animal sanctuary and states, "I don't think I could have someone else do it for me, if I can't do it. If I can't do it, I don't want someone else doing it for me." He then sets out a series of additional reflections about how he should approach eating, which culminate in his decision to become vegan. At this moment Lockwood again describes shedding tears during the film, but this time as a cathartic release that was apparently shared with other audience members.

For Lockwood this emphasis on a particular mode of consumption as a means of combating the vast cowspiracy depicted in the film is not only affectively potent but offers an affirmative course of political action. Yet the film's promotion of a particular way of living as a necessary response to climate change is precisely why it has been criticized in other contexts. Though vegan themselves, environmental activist Danny Chivers is concerned that

> *Cowspiracy* . . . seems to assume that the only people worth targeting with its message are white, Northern and middle-class. One of the most problematic lines in the film is when a commentator says "it's not possible to be a meat-eating environmentalist." This statement is presumably meant to prick the consciences of well-off US eco-activists but it sweeps the struggles of millions of poorer Southern and Indigenous peoples under the carpet. Most of the people fighting for a safer global environment aren't middle-class Northern folks with carbon-heavy lifestyles. They are

the people engaged in frontline battles against fossil fuels, local pollution, and—yes—livestock megafarm projects around the world, and they are leading the way in the defence of our shared climate.[72]

On a superficial level, these debates thus crystallize broader tensions that exist between critical animal studies and situated, relational modes of ethics. *Cowspiracy*'s message seems to evoke precisely the stance that is often criticized, from a more-than-human perspective, for failing to situate eating within a wider environment of interdependent relations. This is precisely the sort of trend that Shotwell identifies as problematic in *Against Purity*, where she sets out her concerns about "purity politics" that suggests ethical solace can be found in consuming the right products in the right way.[73] This politics, Shotwell contends, can reinforce possessive individualism in suggesting that strictly sticking to certain practices of consumption is the key to disentangling individuals from oppressive systems. The flip side of purity politics is that it neglects structural inequalities that might make certain consumption practices inaccessible to particular communities. If a guarantee of purity is associated with certain forms of consumption and not others, ethics becomes attainable only to "those who are most privileged by and within the system."[74] A consequence of this politics is thus that it implicitly positions certain communities as being part of the problem due to their ongoing implication in the systems at stake, precisely the issue identified by Chivers above.

In a sense, therefore, Shotwell's arguments build on Lorimer by pointing to further limitations of an affective logic oriented around shock. However, Shotwell also helpfully complicates matters in drawing attention to an equally pernicious set of problems that persist in arguments typically leveled *against* purity politics (or at least perspectives that are labeled as purity politics in an overly hasty way). For instance, a sort of reverse moralism is often at work in key texts that have criticized activists for their elitism, and critical responses to vegetarian ecofeminism often conclude with assertions that no position is truly innocent and without violence.[75] Despite ultimately arguing for the importance of resisting moral certitude, Shotwell nonetheless argues that an emphasis on the noninnocence of *all* positions is not enough in itself, at least when it comes to finding a means of addressing questions of political action.[76] These arguments resonate with the issues I discussed in previous chapters, which emerged in the context of protest camps and food giveaways, where sometimes the answer as to which approach is overly moralistic and shuts things down, and which "stayed with the trouble," was not obvious.

A "pluralist" stance that appears to be more open, by refusing to preclude particular ways of doing things, can ultimately reproduce existing sociocultural relations in ways that leave hierarchies intact. It should be remembered, therefore, that even pluralism is not neutral and also enacts a particular onto-ethico-epistemological cut, wherein particular realities are brought into being and others foreclosed.

The question of how to decide *which* cuts to make is central to the project set out in *Against Purity*, where Shotwell argues that the only way of negotiating this problem is by finding a way of "holding in view" the systems that lie behind particular ways of eating. It is here that *Cowspiracy* is particularly provocative, because while it does advocate a particular way of eating, it also depicts Anderson going through the *process* of making the decision to do things a certain way, to the exclusion of others, as he reflects on the different possibilities open to him. Anderson's decision, therefore, is not depicted as him imposing an abstract moral framework but is bound up with visceral felt responsibility.[77] Despite the important issues with *Cowspiracy* raised by activists such as Chivers, therefore, it is also important to acknowledge the complex affective politics that arises from its juxtaposition of shocking and sentimental imagery, and the way this approach draws attention to systemic issues (even if the solution to these issues is ultimately resolved too neatly).[78]

The complexity of mediating felt responsibility is approached from a slightly different angle by another activist initiative, iAnimal, which seems to offer a more promising opening for fostering situated ethics.[79] In his recent research project on the campaign, Mike Goodman describes engaging his students with iAnimal, a campaign of short films pitched as "virtual reality" that are produced by the campaigning group Animal Equality.[80] Each "experience" promises a 360-degree perspective on what it feels like to be a farm animal, with videos available on YouTube or—more dramatically—at events where participants are invited to wear virtual-reality headsets for a more immersive experience. The project thus seems to offer a means of overcoming the problem of enacting situated care at a distance, which has made the approaches advocated in nonanthropocentric theoretical contexts difficult to engage in on the part of activists. In light of these potentials, Goodman asks whether iAnimal could open scope for enacting the sort of shared suffering put forward by Haraway, or whether it offers a more complicated reading than this.

An examination of scenes from one of these films, "The Dairy Industry in 360°," suggests that again an analysis is needed that pushes at the films' contradictions. The film opens with a graphic birthing scene, where two workers pull a calf from his mother. Though the scene is clearly designed to shock and elicit an emotional response from audiences, even as the calf's wet body slaps onto the concrete floor actor Evanna Lynch's voice-over underlines the routine nature of such processes by stating, "Just like your mother, a cow has to give birth to give milk."[81] The remainder of the short, five-minute, film is constructed as a biography of the life of the adult cow after separation from her calf, and moves from the peak of her productivity to her death. As in the birthing scene, the camera remains close to the cow's body throughout the film in order to—as Animal Equality's founder suggests—cause participants to "feel trapped, just as the animals are."[82] Although the audience never embodies the cow, the camera, positioned at an angle slightly above her head, still works to mirror her experience; the audience's view is similarly bounded by the narrow confines of the white slaughterhouse pen. Again, the voice-over works to draw out the juxtaposition between the scene's routineness and the sense of shock elicited by Lynch's description of how "her cries of distress echo as she smells the blood of those killed before her." The film climaxes as a worker arrives to euthanize the cow, and as the tension builds, she raises her head level with the camera, making eye contact for a fraction of a section before she collapses.

Despite its focus on bodily engagements, like the beagle images discussed in the previous chapter, the approach taken in iAnimal is ultimately intensely ambiguous in terms of its relationship with embodied care ethics. While each of iAnimal's experiences draws attention to the shared vulnerability of bodies in a broad sense, it is ultimately at odds with the sort of embodied empathy encouraged by thinkers such as Vinciane Despret, for instance. Though iAnimal maintains an emphasis on corporeality as a site of ethical potential, this is where any similarities with the sort of somatic care that has been advocated in theoretical contexts (see previous chapter) begin and end. In "iAnimal Pigs," the film described by Goodman, tensions become still more pronounced as the film takes places from the perspective of a factory-farmed pig, with the experience literally constructing animals as humans-in-fur-coats. The neat mapping of human onto animal is reinforced by the voice-over, where perceptions of what the pig is experiencing are narrated by musician Tony Kanal, and audience-participants are told how to interpret particular sensations and perceptions, with the aim of generating sympathy.[83] Rather than a radical break with anthropocentrism, therefore, the campaign resonates with the sort of "humanist posthumanism" identified by Cary Wolfe, wherein the push to extend ethical frameworks beyond the human is tempered by accomplishing this with frameworks predicated on anthropocentric humanism.[84]

The approach taken by iAnimal thus ultimately seems to reflect exactly the colonizing mode of embodied empathy that Despret is trying to unpick when she argues that partial sensory affinities between bodies (rather than the wholesale mapping evident in this particular enterprise) are the key to fostering understanding and ethical obligation. As described by Goodman, for instance, "iAnimal Pigs" culminates by shifting vantage point, so viewers no longer see the world through the animals' eyes but instead witness the moment of slaughter. This initial mapping of human experience onto (and into) animal bodies, coupled with the final shift in viewpoint, is thus doubly anthropocentric in coupling anthropomorphism with the sharp separation of human from animal.

As was also the case with the beagle rescue images, *Planet Earth II*'s turtles, *Electrocuting an Elephant*, and *Cowspiracy*, however, it is important not to dismiss iAnimal's politics. Like the Green Hill images, it would be easy to dismiss these videos as another brand of anthropomorphism, but the experiences are again far messier than this. Rather than viewing the humanist values that structure iAnimal as straightforwardly undercutting attempts to shift beyond the human, I suggest it is more useful to see these tendencies as lying in generative tension with one another. Even though iAnimal might not elicit as dramatic a legal response as the Green Hill beagle campaign, Goodman made clear that these experiences did elicit a visceral reaction from his students, and this suggests the resonance of the initiative cannot be dismissed easily. Both *Cowspiracy* and iAnimal, therefore, open up a set of difficult questions about the constraints and potentials offered by spectacular environmentalisms, which can be brought into sharper focus when turning to a final figure: the cow-as-worker.

Working cows are provocative because they offer a mirror image of campaigning films; rather than displaying anthropocentric tendencies *despite* efforts to extend ethical frameworks beyond the human, an emphasis on cow labor—though often deployed in ironic and playful ways in marketing contexts—can *inadvertently* offer ways of thinking beyond the human.[85] In the United Kingdom in 2010, for instance, a campaign by butter brand Anchor depicted "cows at work," not on a farm but in a factory. A team of cows donning hard hats and high-visibility jackets were represented as being hard at work (though contented), pausing only to purchase neat squares of turf from the office vending machine.[86] Five years later in Sweden, the scene in another dairy marketing campaign offered a slightly different approach, with cows instead puncturing their working day with mindfulness classes or enjoying hay from organic food trucks.[87] These are just two evocative examples of a popular marketing trope, and, for all these contrasting scenes say about shifts in workplace environments and the rise of the happiness industry, the cow-as-worker is an important figure more broadly.[88]

The question raised by these marketing campaigns is, as Tobias Linné puts it, what does it mean to bring cow labor into the foreground in this context, when so much critical academic work and activism argues that making animals visible is integral to contesting animal consumption? Linné articulates what is at stake in figuring the cow as worker, by asking whether these marketing representations might contain subversive qualities that cut against the grain of the advertisements themselves. As Linné suggests, "when domestic labor remains unrecognized, this masks processes of exploitation within the home. In a similar way, animal labor and exploitation in dairy production remains unrecognized, the cows' subsumption by the logics of the market being masked."[89] Though the appearance of cows on Instagram and Facebook was originally intended as an amusing way to foster social media engagement on the part of dairy companies such as Arla, Linné argues that it is informative to examine more provocative dimensions of these campaigns. Can such images, for instance, inadvertently foreground the constitutive role of animal bodies in processes of value generation, or even create space for grasping whether affinities could be drawn between human and animal workers? These questions of shared labor are becoming ever more urgent in light of a growing body of research that is revisiting Marxist frameworks in order to figure the more-than-human dimensions of labor.[90]

Despite posing these evocative questions about working cows, Linné's own ethnographic study of the Swedish dairy industry (both individually and with Helena Pedersen) ultimately fails to bear out this initial optimism. Although Linné draws attention to the subversive potential of bringing cow bodies to the fore, Linné found that these campaigns illustrated only part of bovine labor: that which articulates neatly with pastoral narratives and sits with what a number of theorists have identified as a "new carnist" or "happy meat" discourse.[91] In Sweden, moreover, these marketing campaigns were just one strategy among many that sought to articulate a story of dairy farming. The campaigns were coupled with national initiatives such as education programs, where children tour farms to see "the way of the milk," as well as open days where members of the public were invited to see cows released onto pasture.[92] These events often culminated in tasting events, where members of the public were encouraged to taste the end products of what they had witnessed. In order to construct a narrative in which palatable aspects of cows' working lives were brought to the fore, other aspects of their lives were made invisible, such as time spent in barns, calving, the processing of milk, the human labor sustaining these processes, and the relationships between small farms and large dairy corporations.

The cow-as-worker, therefore, is helpful in elucidating a particular problematic related to "making visible," which brings things full circle to the recurring problems faced by anti-fast-food activists. As discussed in chapters 1 and 2, attempts by McDonald's activists to reveal the vast sociotechnical networks that enrolled human, nonhuman animal, and environmental actors alike were complicated by the corporation's green rebranding. With Linné's example, this rebranding directly counters activist narratives and complicates a reliance on straightforward practices of visibility, thus, by extension, posing difficulties for the sort of "holding in view" advocated by Shotwell. It is, I suggest, vital to recognize the difficulties of "holding in view" in a context where activists have to work against existing narratives that appear to invalidate their concerns. Indeed, understanding the difficulties of operating in popular cultural environments is essential in better understanding the contradictions of spectacular environmentalisms, and recognizing that they have to negotiate a mediated terrain where—as in the McLibel trial—the resources available to activists and corporate actors are dramatically uneven. In such contexts logics of shock and sentimentalism, or anthropomorphic imagery, might play an important role in making normative relations surrounding food explicit, in order to create space for asking whether and how they could be configured differently. It is important, therefore, not to simplistically label certain approaches as apolitical lifestyle politics, without taking into account the constraints they are working within and against.

Beyond Affective Inequalities

This chapter has foregrounded the danger of categorizing texts according to degrees of anthropomorphism or anthropocentrism, because of the way this line of argument can work to denigrate sentimentality and flatten the messy politics offered by texts that utilize shocking or awesome imagery. Even if such acts of classification were desirable, the texts discussed throughout this chapter illustrate how seemingly straightforward instances of spectacular environmentalisms often resist easy categorization. Yet, even as I have emphasized the productive ambivalence of particular texts, I am not pushing for an uncritical or celebratory stance toward anthropomorphic depictions of animals on-screen. Caution still needs to be taken when particular representations of animals shore up anthropocentric humanist norms, be it through reinscribing human exceptionalism, uncritical anthropomorphism, or hierarchical modes of advocacy. Instead, I have sought to highlight the importance of undertaking a careful analysis of the ways that anthropocentric and nonanthropocentric tendencies can work through one another. Making these entanglements visible, I suggest, is vital in grasping the ambiguous political openings offered (or, crucially, foreclosed) by popular texts.

Figures such as Topsy, for instance, help to elucidate the messy histories of media representations of animals, which make texts difficult to categorize. Even texts that appear to have simplistic, self-evident meanings can offer a diverse range of political openings and, in doing so, unsettle the metonymic slippage that often occurs between anthropomorphism, anthropocentrism, and sentimentality. As elucidated by *Planet Earth II*'s turtles, moreover, it is important to attend to the sorts of obligations opened up by these texts, obligations that can exist in "sentimental" portrayals as well as more experimental media. Dismissing the sentimental out of hand does justice to neither its ambivalent qualities nor to the complexity of any potential ethico-political responses on the part of those who engage with such texts. As with the social media campaigns outlined in the previous chapter, moreover, the political purchase gained by using shocking or sentimental imagery can make it difficult to dismiss, despite all of its problems.

With these arguments in mind, it nonetheless remains important to be aware of the uneasy relationship between certain texts (such as *Cowspiracy* and iAnimal) and the hierarchical tendencies of purity politics. If the focus is wholly on consumer agency then this inevitably neglects structural inequalities, inequalities that both prop up particular sociotechnical systems and make consumer activism against these systems difficult to realize for certain communities. The specter of purity politics does, of course, haunt activism at different scales, and the struggles of activists described in previous chapters help testify to how this problem is negotiated in contexts such as grassroots protest camps and pamphleteering campaigns, rather than being a problem faced solely by nongovernmental organizations such as Animal Equality. The problem of ethical lifestylism, however, becomes particularly pronounced in campaigning contexts that are detached from the sorts of responsibility practices engaged in at a grassroots level.

Spectacular environmentalisms, therefore, both speak to the problems of articulation outlined in the first chapter and add another dimension to these issues. Just as it is dangerous to insist on a particular model of articulation, because this ignores sociotechnical factors that might prevent particular groups of people from engaging in this approach, it is also dangerous to privilege certain representations of animals (such as those that eschew the homely and sentimental in favor of depicting animal alterity). This move not only dismisses what is productive and politically significant about these texts but, paradoxically, can make it more difficult to gain critical purchase on them.

While it is critically important not to excuse recourse to purity politics, in practice sentiment that is often labelled "purist" usually exists as one strand of spectacular environmentalisms that jostles with competing tendencies. Perhaps, therefore, where the texts discussed here are the most productive is in elucidating the points where an emphasis on micropolitical lifestyle does important work, and the points where it is in danger of undercutting broader intersectional struggles. The subversive potential of emphasizing intersections between human and animal labor holds particular promise in this regard, if taken up more decisively in awareness-raising contexts, rather than leaving these narratives to be co-opted by marketing campaigns. Indeed, as chapter 3 shows, there is already a strong tradition of tracing connections between human and animal work in grassroots activist contexts, which could be harnessed to political ends within popular awareness campaigns.

In order to create the necessary space for asking critical questions about popular media representations of animals, therefore, it is not enough to talk in general terms about the way texts inhabit particular logics. Instead, to frame things again in relation to feminist media studies, it is important to pay attention to the specificities of texts. In particular, there is a need to attend to specific moments where anthropocentric tendencies are entangled with radical imagery in ways that undercut more profound gestures toward structural change. It is important to ask, in other words, not just whether approaches to environmental politics that *seem* to be untroubling to human exceptionalism are sometimes more unsettling than they appear, but how the troublemaking capacities of these forms of politics could be enhanced.

Conclusion

An Ethics of Exclusion

In posing the question of what comes after entanglement, this book has not aimed to deny, erase, or simplify the complex relations that exist among the disparate species, technologies, and practices that compose lived reality. What I *have* sought to create space for are questions about which courses of action are available once these entanglements have been recognized. As elucidated throughout the book, complex issues demand complex ethico-political responses, and nothing has an easy solution: from the micropolitics of eating or waste disposal, to questions about how to intervene in large-scale systems associated with animal agriculture or technoscientific practice. At the same time, it is still important to find ways of acting amid this complexity. In order to even begin to address the difficult set of questions that surround action and intervention, however, there needs to be a shift in conceptual focus from an ethics of born of entanglement toward an ethics of exclusion.

What I have pushed for in making this argument is twofold: first, a change in *emphasis* that engages with the productive role of exclusion. Rather than focusing on the complex multispecies entanglements that compose lived reality, it is also important to grasp the constitutive role of exclusion. Exclusion in this sense is not just a negation of something but plays a necessary and creative role in the fabric of the world.

This understanding of exclusion as something with creative force feeds into my second, related argument: it is essential not just to acknowledge the constitutive role of exclusion (though this is important). It is also vital to address the more fundamental implications of this recognition by actively *politicizing* exclusion. Exclusion is not just something that happens, an inevitable component of the ever-evolving, entangled composition of the world, but is often bound up with particular sociotechnical infrastructures and political decisions. It is vital, therefore, to find ways of taking responsibility for the exclusions that are fostered by specific entanglements. Indeed, this form of responsibility is necessary in creating space for future transformation, by making exclusions visible and open to contestation by those who are most affected by them.[1] In the final pages of this book, I flesh out this argument in more depth, in order to articulate the value of exclusion as an ethical orientation.

Exclusion as Constitutive and Creative

The constitutive role of exclusion is already acknowledged within relational, more-than-human approaches; indeed, it is critically important. To reiterate arguments made in chapter 2, for instance, Karen Barad's conception of agential cuts sees the performative production of matter as a complementary process, wherein the worlds manifested through particular assemblages necessarily occur at the expense of other possibilities. In doing so, she calls for attention to be paid to the boundary-making practices through which these cuts are enacted and offers a reminder that things could always be otherwise if this assemblage was composed and performed in a different way.[2] Joanna Latimer pushes a Baradian approach still further in foregrounding the lack of neatness in any cut, conceiving boundary making instead as more complex thresholds through which particular entanglements unravel as others are brought into being.[3] These thresholds are sites of difficult ethical engagements and decisions, where certain values, ways of being, and even lives are prioritized over others.

As I have argued with Gregory Hollin, Isla Forsyth, and Tracey Potts, however, despite the ethical potential of focusing on these cuts and thresholds, this theme is frequently treated as *secondary* to relationality.[4] Intra-active, performative accounts of the material world are often used to denaturalize hierarchical distinctions between different actors in order to underline the notion that even matter itself can be otherwise. While such approaches might be important in unsettling anthropocentric or ethnocentric relations and classifications, the emphasis remains on what is brought into being. This effectively positions the constitutive role of exclusion as a component or consequence of particular relations emerging.

In theoretical work that insists on the relation as the smallest unit of analysis, the ethical emphasis remains on the potential inherent in encounters, relations, and comings-together rather than on the boundary-making practices that instantiate them. As Elizabeth Wilson puts it, therefore, there is a need to "find a way to articulate more fully what Barad gestures towards but seems unable to entirely countenance: that negativity, never under our control, has a permanent place in the spacetimemattering of the world."[5] This argument, however, can itself be pushed further. Although certain exclusions might be beyond control, in other contexts anthropogenic activities or inequalities fostered by the behavior of corporate actors have been afforded heightened influence, and sustained responsibility needs to be taken for the relations that they constitute. It is in such contexts that exclusion can be understood in more political terms.

Although nothing might exist outside of relation, *certain* things might need distance from *certain* relations in order to allow particular realities to be enacted and preserved—and creating this distance is a decisively ethico-political concern. As Thom van Dooren elegantly argues, in the context of Hawaiian crow conservation, "we don't need to buy into a simplistic nature/culture dualism to believe that some creatures, some places, would be better off in a range of different ways if we carefully and deliberately limited our involvement with them."[6] Resisting essentialist understandings of what an authentic crow might be, van Dooren instead traces how human activities have fundamentally altered the assemblage through which crow identity is materialized. Certain human ways of living, in other words, have impinged on the relations through which a particular performance of crow identity is realized: a performance that is not individual but has been instantiated through the intergenerational work of the species.

The stakes of these developments are significant in that they have not just resulted in a different materialization of what it means to be a crow, but one characterized by "irreparable" change in crow capacities that might make future survival difficult.[7] What is brought home by this example is not just that it is important to maintain distance between certain actors in certain contexts, but that particular realities are materially (and perhaps irrevocably) foreclosed if other relations are brought into being. Different ways of being are often mutually exclusive, and responsibility needs to be taken for which world is materialized. Where van Dooren's sympathetic criticism of certain modes of relating is especially informative, therefore, is in relation to questions of action and intervention. What is hinted at by van Dooren is the need to preserve some sense of how deliberate acts of distancing and exclusion can play an important role in fostering multispecies flourishing.

This line of argument can be elaborated still further by turning back to Franklin Ginn's work on more everyday negotiations with garden slugs (see the introductory chapter). For Ginn, there is a need to address the productive role of distance and exclusion, something that, he argues, is currently a "considerable blind spot of an affirmative ethics based on meeting, matter and bodies-in-relation." In order to instead underline the constitutive role of exclusion, he suggests a step forward could be through "acknowledging the ontology, not of relation, but of relation/detachment" and of the complex ethics in practices "such as distancing, spacing, hiding or retreating."[8] Recognizing the productive work of particular exclusions, Ginn suggests, could be a means of realizing a "new ethics of detachment that could work sometimes around, sometimes in parallel, sometimes antecedent to or after, the relation and practices of relating."[9] Elaborating on these arguments, I argue that it is necessary to understand not just how an ethics oriented around exclusion could work alongside relationality, but how such an ethics offers provocations for a conceptual emphasis on relations.

Emphasizing Exclusion

Throughout the book the implications of exclusion for relational ethics have become clear, in part, through emphasizing its constitutive role in slightly different activist environments. Through attending to tensions between particular strands of theory and practice throughout the book, I have worked to acknowledge and emphasize the role of exclusion. Each chapter has built up a multilayered picture of exclusions that have emerged in a range of political contexts. The tactical interventions engaged in by the activist groups I turned to, for instance, were often actively contesting particular exclusions.

These processes of contestation were evident in a very concrete sense during the McLibel trial (see chapter 1), when activists focused on mundane points of friction within McDonald's infrastructures—from blocked drains to broken fryers—in order to denaturalize the reality effects of these infrastructures, to put things in Annemarie Mol's terms.[10] Disrupting the illusion of smooth efficiency was, in turn, a means of disrupting the inevitability of particular relations and ways of being, and this opened space to contest the particular ways that publics, workers, and animals were enrolled by McDonald's infrastructures. The tactics engaged in by activists were thus suggestive of how particular exclusions could be made to matter in both political and ethical terms, through being rendered visible and open to contestation.

The McLibel tactics, then, reveal two distinct *forms* of exclusion. In part, this instance of activism points to broader questions of framing and the need, in some situations, to contest the boundary of particular frames, that is, boundaries that determine who is perceived as a subject in a given situation, or whose opinion is counted. As argued in other chapters throughout the book, even contradictory approaches (such as spectacular environmentalism or emotive imagery) can work to bring those who are routinely dismissed from ethical consideration back into the frame. As with debates around environmentalist and anti-speciesist consumption practices, though, care must be taken to ensure that the methods used to contest particular oppressions do not reinscribe other racialized, gendered, and classed inequalities. Indeed, the danger of inadvertently shoring up other forms of oppression is precisely why it is important to foster responsibility for any course of action (and its attendant exclusions).

Questions of who is or is not included in ethical framings, however, are not the only form of exclusion that is hinted at by activist practice. As I have traced tensions between instances of theory and practice throughout the book, what has also come to the fore is exclusion in a slightly different sense. This meaning of *exclusion* draws on the feminist materialist lineage described above and refers to something constitutive that plays a role in actively materializing particular ways of being. To revisit the McLibel trial: in disrupting the particular relations bound up with McDonald's, the activists were not just seeking to carve out space for alternative voices to be heard, or particular forms of agency to be recognized (though this was an important part of their practice). The campaign was also working to disrupt the relations through which oppressive realities were *materialized*, be it in the setting of the courtroom or in the context of McDonald's own sociotechnical arrangements.

These acts of contestation are where things become especially messy, as exclusion in a material sense is not just something rendered visible or contested by activist practice; this form of exclusion is also an integral *part* of political intervention. Instances of activism in all of the chapters have, in a sense, been drawing attention to particular forms of social organization and ways of living with other species, which activists believed were being constitutively excluded from existing relations. In order to create space for these alternatives to emerge it was thus often necessary to contest, or exclude, the relations that prevented them from coming into being. Exclusion in a material sense, therefore, is not intrinsically negative or something that shuts down agency, but more akin to the affirmative politics of distancing and alterity evoked by van Dooren and Ginn. In explicitly opposing capitalist or patriarchal relations, for example, the activists discussed in the first half of the book were working to contest the exclusions fostered by these arrangements but, in doing so, necessarily made exclusions of their own by pushing for explicitly anticapitalist alternatives at the expense of other ways of being.

What becomes apparent when bringing exclusion's constitutive role to the fore in explicitly political contexts, therefore, is that its ethics lies not just in matters of classification and where to draw the line in relation to who is or isn't part of particular ethical community. It is not just that certain forms of agency, practices, or realities are marginalized by given sociotechnical arrangements, but that they cannot even come into *being* when other relations exist.[11] From this perspective, interventional forms of activism that adopt staunch ethical stands are given a slightly different resonance.

The act of opposing or contesting particular relations is necessarily an act of exclusion, but this does not straightforwardly equate to shutting down ways of being or imposing totalizing moral stands. Instead, in certain contexts, acts of contestation and distancing are precisely what clear space for alternative realities and expressions of agency to emerge. The activist groups discussed throughout the book, for instance, often made purposeful decisions to oppose particular relations, be this through activities such as skill shares that were designed to oppose technocratic hierarchies, or the contestation of particular systems that linked humans and other animals together in oppressive ways. These interventions were often vital in revealing and overcoming hierarchical and exploitative relations that reduced others' agency.

Recognizing that every course of action carries attendant exclusions is important, therefore, in complicating notions about what modes of ethics are necessary in responding to entangled worlds. At the same time, it is necessary to move beyond simply acknowledging the inevitable role of exclusion, as this could prove as paralyzing for questions of action and intervention as recognizing that everything is entangled. Exclusions, I argue, do not just need to be acknowledged but politicized. In order to open space for more political questions to be asked, however, it is necessary to negotiate the exclusions that—perhaps paradoxically—constitute relationality itself.

Politicizing Exclusion

In *Staying with the Trouble*, Donna Haraway reiterates a refrain that has been central to her work: "Who renders whom capable of what, and at what price, borne by whom?"[12] What I have elucidated throughout this book is that it is difficult to really get at these questions within a relational framework predicated on the recognition of entanglement and complexity. This is not to say that an emphasis on complex multispecies interdependencies is not hugely valuable; what was so radical about Haraway's conception of companion species was the way it figured the relation as the fundamental "unit of analysis."[13] In placing the emphasis on irreducible complexity and co-becoming—"turtles all the way down"—Haraway and related thinkers have been able to decenter the human in critically important ways.[14] As Stacy Alaimo puts it, an emphasis on interdependencies and entanglements generates questions about "what forms of ethics and politics arise from the sense of being embedded in, exposed to, and even composed of the very stuff of a rapidly transforming material world."[15]

What is implied by these arguments, however, is not only that the relation is the fundamental unit of analysis but that ethics can only emerge *from* relation (or at least an ethics that is sufficiently nuanced and responsive). The consequence of this argument is that it places the ethical emphasis on coming together and proximity in ways that have significant consequence for the other modes of ethics or political practice that are necessarily foreclosed by this emphasis. While certain ethical potentials are opened up, others are shut down. To reiterate Maria Puig de la Bellacasa's argument, the way that relational approaches redistribute agency can sometimes make it *more* difficult to realize "ethico-political commitment and obligations," rather than opening up these obligations in the way that is often suggested.[16]

A focus on specific relations or encounters in themselves, for instance, can mask asymmetrical distributions of agency that not only constrain what ways of being are possible in a given situation but, in doing so, inhibit possibilities for future transformation.[17] This problem has resurfaced throughout the book, in a range of contexts. For example, in alternative media networks and protest camps, informal hierarchies were sometimes masked by the assumption that particular ways of organizing, or certain technologies, secured openness and transparency (see chapters 2 and 3). The invisibility of these hierarchies masked the reproduction of normative, intersectional inequalities that made it difficult for certain actors to express agency. Indeed, some of the most significant work activists engaged in was finding creative ways of rendering these hierarchies visible in order to open them to challenge or—even more productively—engaging in practices designed to spread expertise.

Similar problems of informal or at least invisible hierarchies emerged in the context of multispecies relations. As outlined in chapters 4 and 5, longer histories of human-animal relations (such as those with primates and laboratory beagles) or relations between particular epistemic communities (such as experts and publics) can work to foreclose disruptive expressions of agency in advance of the relations and encounters themselves. What these examples have shown is that sometimes it might *seem* like space is being created for certain actors to impose their obligations, or for transformative expressions of agency and resistance to manifest themselves, when these possibilities have already been rendered impossible through prior encounters and inequalities.

These difficulties do not just pose practical challenges for activists but have broader theoretical implications and are conceptually significant when it comes to matters of responsibility. What I have argued throughout the book is that entangled, relational visions of the world offer sparse means of taking responsibility for their own exclusions. By their very nature, these approaches seem to accommodate a multitude of different ways of being and encompass irreducible complexity: but this is precisely the problem. Due to these approaches appearing to be open to difference and plurality, their own exclusions are often difficult to detect; this is dangerous because in some instances relational modes of ethics segue with precisely the structures of domination they are trying to contest. The first chapter's discussion of McLibel, for example, traced how the asymmetries activists were negotiating during the trial reproduced broader social inequalities. In this context, adherence to the modes of articulation pushed for in theoretical contexts made it difficult to speak at all, let alone to actively contest sociotechnical norms. Similarly, the latter section of the book traced how distinctions between types of care that have been prominent in theoretical contexts (wherein abstract animal rights frameworks are seen as too neat, while embodied care toward animals is seen to secure ongoing responsibility) can map onto hierarchies of expertise constructed in mass-media contexts.

What has come to the fore throughout the book, therefore, is that it is not enough to acknowledge the noninnocence of all forms of relation. This recognition can inadvertently naturalize the exclusions that are constitutive not just of particular relations but of relational ethics itself, rather than promoting ownership of these exclusions. An emphasis on relational *ontologies* automatically labels thought and action that question particular relations as somehow denying the more-than-human composition of the world.[18] The foreclosure of particular forms of criticism is thus naturalized, because critical-activist responses are positioned as being simply at odds with material reality. This form of argumentation legitimizes the exclusion of all forms of ethics that are labeled in this way on the grounds that they fail to recognize how things "really" are.

The difficulty of engaging in even sympathetic criticism of relational theoretical work thus has evocative parallels with Jo Freeman's arguments about structureless political groups. For Freeman, the difficulty is that attempts to reveal informal hierarchies often lead to the very group members who draw attention to these issues being accused of pulling rank and attempting to reinstate hierarchical structures. This distribution of blame effectively renders those who are the most marginalized as themselves being the problem. What Freeman highlights is how openness and structurelessness—values that are presented as creating space for diversity and dissent—can both obscure the persistence of hierarchies and place these hierarchies beyond criticism. These dynamics mean that intersectional inequalities and exclusions are reinscribed and naturalized with no possibility of challenging them—or, at least, no way of challenging them that can be accommodated within a structureless ethico-political approach.

For Freeman, a means of overcoming these informal hierarchies is by recognizing that in some instances temporary, contingent structures are necessary in order to actively oppose particular ways of being and create space for alternatives to emerge.[19] These arguments, I suggest, need to be read back against contemporary theoretical work, as it is precisely the sorts of contingent structures identified by Freeman—those that oppose or contest particular relations and that are vital in clearing the space for alternatives—that are often incommensurable with relational modes of ethics. In other words, relationality is often constituted by foreclosing messy forms of criticism and intervention that are essential for enacting ethical responsibility in practice.

It is in creating room for these productive forms of responsibility that exclusions offer especially productive political purchase. Taking exclusion, rather than entanglement, as the key site of ethico-political importance changes the sorts of questions and issues that need to be prioritized. Instead of finding ways of responding to and respecting complexity, what becomes important is taking responsibility for exclusion. As argued throughout the book, the project of realizing responsibility demands a more expansive, heterogeneous, and perhaps messier set of responses that go beyond an ethics oriented around proximal relations and encounters, to instead force attention to longer histories and intersectional inequalities that inform these relations. In some contexts this might make it necessary to (critically) recuperate practices that are ordinarily sidelined from conceptual consideration (such as rights or emotional responses that are often sidelined for being overly sentimental).

Expanding the Realm of Ethical Possibilities

By offering a thicker account of activist tactics that, on a superficial level, appear to be at odds with calls for more situated and relational modes of ethics, I have worked to combat the inadvertent marginalization of productive modes of action, intervention, and accountability. In order to recuperate some of these tactics, it has been necessary to firmly situate the work of activism as itself operating in distinct ecologies and working with and against preexisting sociotechnical arrangements. Through focusing on how activists have tactically negotiated constraints on their practice, I have sought to account for how and why tactics that appear to adopt totalizing stances are engaged with in particular sociohistorical settings: from decisive practices such as veganism to uses of anthropomorphic imagery within spectacular environmentalisms. By situating activism in this way, I have not unreflexively justified these or related tactics but instead offered a sense of how they are manifested and what work they accomplish in particular contexts.

The tactical interventions engaged with throughout the book have elucidated a broad range of practices for making a difference in the composition of the world, by contesting the relations that foreclose particular forms of agency. The problem is that, as argued above, practices that have often proved valuable in practice for fostering responsibility are often inadvertently foreclosed by relational, more-than-human approaches. What the examples discussed throughout the book have illustrated—from alternative media infrastructures to friendly encounters with laboratory beagles—is that ethical and epistemological responsibility is not found solely in the moment of encounter itself. Indeed, valorizing these moments and relations can obscure rather than open up responsibility. It is instead important to constantly ask who or what is being excluded when certain realities are materialized at the expense of others, to find ways of taking responsibility for these exclusions, and in some instances to contest them.

These arguments have implications for the way action and intervention are often conceived. As I have suggested here, contemporary theoretical debates are routinely characterized by the assumption that certain approaches offer ongoing responsibility toward entangled worlds while other—more totalizing or essentialist—approaches shut down this responsibility. Emphasizing and politicizing exclusion helps to complicate such a narrative.

In adopting a focus on activist practice, I have shown what is at stake, in epistemological and ethical terms, when particular tactics or strands of critical thought are excluded from debate in ways that naturalize these exclusions. This point, moreover, has broader significance to relational approaches that goes beyond a specific focus on activism. What these instances of activism do is highlight the wider political potentials that can become excluded from conceptual consideration when certain approaches to contesting hierarchy and anthropocentrism become normative. It is dangerous to create a normative sense of what a politics that responds to entanglement and complexity should look like, as any approach—even one that is apparently open and pluralistic—has constitutive exclusions.[20] The problem is when these exclusions are masked and placed beyond contestation: this is what forecloses the potential for future transformation.[21] Yet this naturalization of exclusion is precisely what happens when approaches that appear to be open and responsive fail to take into account preexisting structural inequalities that have already distributed agency in ways that make it difficult for particular actors to contest the relations they are embroiled in.

To overcome uneven distributions of agency, it is not just a matter of recognizing, or placing greater emphasis on, exclusion. The frictions between critical-activist approaches and theoretical work that I have traced throughout the book show that the implications of centering exclusion are more profound than they first appear. A politicized account of exclusion does not necessarily complement relational approaches, in other words, but can productively trouble the way this ethics is constituted.

Perhaps, then, as I hinted at in the introduction, asking what comes after entanglement is the wrong framing of the question. Exclusion does not necessarily just come after, work around, or give birth to relations; sometimes its ethical potential is precisely in the purposeful way it destroys particular entanglements in order to create space for alternatives.[22] Understanding exclusion not only as something that is inevitable and constitutive but also as a key site where agency is distributed, and where responsibility needs to be taken, necessarily shifts the types of questions that need to be addressed. Instead of finding ways of respecting and responding to complexity, it is important to ask how to be more accountable to the exclusions that are inevitably fostered by any course of action (or indeed inaction).[23] Rather than seeing the relation itself as the foundation of ethical accountability, in other words, meaningful responsibility can only be taken by centralizing and politicizing the exclusions that have brought these relations into being.

Notes

Introduction

1 Donna Haraway, *When Species Meet* (Minneapolis: University of Minnesota Press, 2008). Haraway's opening arguments in this text are a key point of reference for these developments.

2 For example, Kelsi Nagy and Phillip David Johnson, *Trash Animals: How We Live with Nature's Filthy, Feral, Invasive, and Unwanted Species* (Minneapolis: University of Minnesota Press, 2013); and Jonathan L. Clark, "Uncharismatic Invasives," *Environmental Humanities* 6, no. 1 (2015): 29–52.

3 The emphasis here should be on the word *particular*; as Juanita Sundberg argues, it is important to recognize the colonizing move that is made when certain theoretical perspectives present themselves as reacting against dualistic thought. In the very process of criticizing dualisms, this line of argument can inadvertently present dualistic thinking as the universal ground it is reacting against. As Sundberg points out, such claims both neglect diverse nondualistic ways of thinking that have always existed (especially outside of an Anglo-European tradition) and also obscure the debt that contemporary nondualistic thought owes to Indigenous epistemologies and ontologies. This argument is also revisited in the next chapter. See Juanita Sundberg, "Decolonizing Posthumanist Geographies," *Cultural Geographies* 21, no. 1 (2014): 33–47.

4 For an overview of the hopes attached to relational, more-than-human ethics, see Eva Giraud et al., "A Feminist Menagerie," *Feminist Review* 118, no. 1 (2018): 61–79.

5 For an example of how relational ethics has been put to work as a means of cultivating responsibilities toward multispecies and more-than-human communities, see Deborah Bird Rose, Thom van Dooren, and Matthew Chrulew, eds. *Extinction Studies: Stories of Time, Death, and Generations* (New York: Columbia University Press, 2017).

6 For analyses of the complexities of plastic, see Stacy Alaimo, *Exposed: Environmental Politics and Pleasures in Posthuman Times* (Minneapolis: University of Minnesota Press, 2016); Deirdre McKay, "Subversive Plasticity," with Padmapani Perez, Ruel Bimuyag, and Raja Shanti Bonnevie, in *The Social Life of Materials: Studies in Materials and Society*, ed. Suzanne Küchler and Adam Drazin (London: Bloomsbury Academic, 2015), 175–192; and Alison Hulme, *On the Commodity Trail: The Journey of a Bargain Store Product from East to West* (London: Bloomsbury, 2015).

7 Thom van Dooren's *Flight Ways: Life and Loss at the Edge of Extinction* offers an instance of the complexity of extinction, tracing how the sharp decline of vultures in India—though in part due to the use of a particular agricultural chemical—needs to be situated in relation to longer histories of multispecies relations as well as social justice issues, such as the impact of poverty on farming practices. Thom van Dooren, *Flight Ways: Life and Loss at the Edge of Extinction* (New York: Columbia University Press, 2014), 45–62.

8 A helpful critique of this form of moralism can be found in Alexis Shotwell, *Against Purity: Living Ethically in Compromised Times* (Minneapolis: University of Minnesota Press, 2016).

9 In her otherwise-sympathetic appraisal of relational ethics, Shotwell notes that such approaches remain unsatisfactory with regard to their ability to make ethical differentiations between distinct types of relations; this argument is explored in more depth in chapter 1. Shotwell, *Against Purity*, 117.

10 Michelle Murphy makes this point about complexity as dispersing responsibility in *Sick Building Syndrome and the Problem of Uncertainty* (Durham, NC: Duke University Press, 2006), 149; I pick up this argument in more depth in chapter 3.

11 *Intervention* here is used in a generic sense of attempting to intervene in a particular political situation; I unpack my understanding of the term toward the end of this introductory chapter in relation to activist tactics. The concept has a more specific meaning in science studies that I do not explicitly draw on here; see Teun Zuiderent-Jerak and Casper Bruun Jensen, "Editorial Introduction: Unpacking 'Intervention' in Science and Technology Studies," *Science as Culture* 16, no. 3 (2007): 227–235.

12 This notion of an ethics of exclusion is derived from collaborative work I engaged in with colleagues, when reflecting on the contributions Karen Barad's work has made to science and technology studies. See Gregory Hollin et al., "(Dis)entangling Barad: Materialisms and Ethics," *Social Studies of Science* 47, no. 6 (2017): 918–941. For further elaboration on what it might mean to centralize the notion of exclusion, as worked through in the context of diagnosis, see Gregory Hollin, "Failing, Hacking, Passing: Autism, Entanglement, and the Ethics of Transformation," *BioSocieties* 12, no. 4 (2017): 611–633.

13 Susan Leigh Star, "Power, Technology and the Phenomenology of Conventions: On Being Allergic to Onions," in *A Sociology of Monsters: Essays on Power, Technology and Domination*, ed. John Law (London: Routledge, 1991), 26–56.

14 For a critical engagement with these developments, see Claire Colebrook, *Death of the PostHuman: Essays on Extinction*, vol. 1 (Ann Arbor, MI: Open University Press, 2014).

15 "We Have Never Been Human" is the title of the first section of Haraway's *When Species Meet* (3–160).

16 This sort of ontological claim is found, for instance, in Haraway's discussion of sympoiesis in *Staying with the Trouble: Making Kin in the Chthulucene* (Durham, NC: Duke University Press, 2016), 58–98.

17 A shift away from hybridity is argued for explicitly in Jamie Lorimer, *Wildlife in the Anthropocene: Conservation after Nature* (Minneapolis: University of Minnesota Press, 2015), 17.

18 The most famous figuration is perhaps Haraway's cyborg. Donna Haraway, "A Cyborg Manifesto: Science, Technology, and Socialist-Feminism in the Late Twentieth Century," in *Simians, Cyborgs, and Women* (London: Routledge, 1991), 127–148. For a contemporary elucidation of the power of figurations, see Michelle Bastian, "Fatally Confused: Telling the Time in the Midst of Ecological Crises," *Environmental Philosophy* 9, no. 1 (2012): 23–48.

19 Karen Barad, *Meeting the Universe Halfway: Quantum Physics and the Entanglement of Matter and Meaning* (Durham, NC: Duke University Press, 2007).

20 Bruno Latour, *We Have Never Been Modern* (Cambridge, MA: Harvard University Press, 1993); and Haraway, *When Species Meet.*

21 For instance, though primarily engaging with different traditions of thought, the argument that underpins texts such as Cynthia Willett's *Interspecies Ethics* and Lori Gruen's *Entangled Empathy* is that the recognition of entanglements between human and nonhuman animals can give rise to less anthropocentric forms of ethics and politics. As I argue throughout the book, these are just two instances of texts that adopt this line of argument. Cynthia Willett, *Interspecies Ethics* (New York: Columbia University Press, 2014); and Lori Gruen, *Entangled Empathy: An Alternative Ethic for Our Relationships with Animals* (Brooklyn, NY: Lantern Books, 2015).

22 The clear point of reference here is the emerging body of work focused on articulating ethico-political responses to the Anthropocene, an era in which human action impacts all life on the planet. See Jamie Lorimer, "Multinatural Geographies for the Anthropocene," *Progress in Human Geography* 36, no. 5 (2012): 593–612.

23 This argument underpins Rosi Braidotti's arguments in *The Posthuman*, especially her claim that a nonanthropocentric theory is vital in providing a conceptual challenge to climate change and sharp declines in biodiversity. Rosi Braidotti, *The Posthuman* (London: Polity, 2012).

24 See, in relation to conservation politics, van Dooren, *Flight Ways*; neuroscience: Felicity Callard and Des Fitzgerald, *Rethinking Interdisciplinarity across the Social Sciences and Neurosciences* (Basingstoke, UK: Palgrave Macmillan, 2016); fine art: Deborah Frizzell and Harry J. Weil, curators, *Women in the Wilderness*, exhibition, Wave Hill, Glyndor Gallery, New York, April 9–July 9, 2017; and quantum physics: Barad, *Meeting the Universe Halfway.*

25 An argument about the lack of space for intervention in the context of posthumanism is made persuasively by Helena Pedersen (although this argument has broader relevance to relational, more-than-human theories), "Release the Moths: Critical Animal Studies and the Posthumanist Impulse," *Culture, Theory and Critique* 52, no. 1 (2011): 65–81.

26 Shotwell, *Against Purity*, 117. It is important to reiterate that this is a highly sympathetic critique of relational frameworks, which maintains a clear sense of their

value as well as suggesting that a more concrete idea is needed about how to respond to the ethics they push for.

27 For an explanation of why activism can inadvertently reinforce humanist norms by placing humans as privileged advocates for nonhumans, see Donna Haraway, "Species Matters, Humane Advocacy: In the Promising Grip of Earthly Oxymorons," in *Species Matters: Humane Advocacy and Cultural Theory*, ed. Marianne DeKoven and Michael Lundblad (New York: Columbia University Press, 2011), 17–26. These specific criticisms of animal rights activism have since been tempered by Haraway in a recent interview with Sarah Franklin, but the underlying suspicion of rights as a framework has been retained. See Sarah Franklin, "Staying with the Manifesto: An Interview with Donna Haraway," *Theory, Culture and Society* 34, no. 4 (2017): 49–63.

28 Haraway, *When Species Meet*, 67.

29 The distinction between killing and "making killable" was articulated in Haraway, *When Species Meet*, 80; the latter refers to the process of rendering a particular life-form as legitimate to kill without ethical reflection. For Haraway, making killable is more ethically problematic than the act of killing other species in itself, and her sympathetic criticisms of activism often hinge on the way that certain tactics might oppose the killing of specific species (often those easiest for humans to relate to) without unsettling the logics and systems that render other species *killable*.

30 Shotwell, for instance, though sympathetic to vegan politics (and vegan herself), criticizes the specific way that certain organizations have tried to promote veganism, arguing that PETA, for instance, is "an organization that . . . has done more than any other, through a series of video and performance interventions, to convince people that vegetarians and vegans are clueless racists unable to take a feminist stance on body politics." Shotwell, *Against Purity*, 121.

31 Lorimer notes that the difficulties posed by life-forms that are harmful are a particular problem for relational, more-than-human approaches to ethics. Lorimer, "Multinatural Geographies." This issue is also noted by Uli Beisel, "Jumping Hurdles with Mosquitos," *Environment and Planning D: Society and Space* 28 (2010): 46–49. It is this difficulty of where to draw the line that Cary Wolfe focuses on in *Before the Law: Humans and Other Animals in a Biopolitical Frame* (Chicago: University of Chicago Press, 2012).

32 Lorimer, "Multinatural Geographies," 604.

33 It is important to note, moreover, that relational approaches are not the sole purveyors of these criticisms; indeed, the frequent positioning of these fields as distinct in making this argument has been accused of masking other traditions of thought that have never been grounded in distinctions between nature and culture. See again Sundberg, "Decolonizing Posthumanist Geographies." For important criticisms from Indigenous scholars, see Zoe Todd, "An Indigenous Feminist's Take on the Ontological Turn: 'Ontology' Is Just Another Word for Colonialism," *Journal of Historical Sociology* 29, no. 1 (2016): 4–22; and Kim TallBear, "Beyond the Life/Not Life Binary: A Feminist-Indigenous Reading of Cryopreservation, Interspecies Thinking, and the New Materialisms," in *Cryopolitics: Frozen Life in a Melting World*, ed. Joanna Radin and Emma Kowal (Cambridge, MA: MIT Press, 2017), 179–200.

34 Bruno Latour, "Why Has Critique Run Out of Steam? From Matters of Fact to Matters of Concern," *Critical Inquiry* 30, no. 2 (2004): 225–248. I elaborate on debates surrounding Latour's "critique of critique" in chapter 4. For a polemical argument for the need for a more critical field of animal studies, see Steven Best, "The Rise of Critical Animal Studies: Putting Theory into Action and Animal Liberation into Higher Education," *Journal for Critical Animal Studies* 7, no. 1 (2009): 9–52.

35 The relationship between critical and mainstream animal studies is fleshed out in chapter 3. To sketch out some important arguments, though, work in CAS is committed to principles of social justice in relation to animals. As with Best's "The Rise of Critical Animal Studies," this political stance can manifest itself as a wholesale criticism of theory (particularly work from a poststructuralist lineage). Other work has adopted a more sympathetic relationship with theoretical work and has instead pushed on conceptual contradictions and tensions within relational ethics. Helena Pedersen warns, for example, that "rather than disturbing species boundaries," an uncritical celebration of human-animal relations "does a colonial work of reinscribing them." Zipporah Weisberg, similarly, criticizes theories of companion species for colluding with "structures of domination" by undermining any capacity to enact a structural critique of speciesism, and Carol J. Adams suggests that practices explored by Haraway (such as hunting and animal breeding) intrinsically position animals as killable resources in precisely the manner that Haraway otherwise condemns. Other thinkers have adopted approaches with a clear affinity for feminist science studies, working critically with more-than-human theoretical work: as in Erika Cudworth's use of posthumanism to develop a complex intersectional critique of human and animal oppressions, or Tom Tyler's critical attempts to move beyond anthropocentric philosophy. See Pedersen, "Release the Moths," 72–73; Zipporah Weisberg, "The Broken Promises of Monsters: Haraway, Animals and the Humanist Legacy," *Journal for Critical Animal Studies* 7, no. 2 (2009): 22; Carol J. Adams, "An Animal Manifesto: Gender, Identity, and Vegan-Feminism in the Twenty-First Century," interview by Tom Tyler, *Parallax* 12, no. 1 (2006): 120–128; Erika Cudworth, *Social Lives with Other Animals: Tales of Sex, Death and Love* (Basingstoke, UK: Palgrave Macmillan, 2011); and Tom Tyler, *CIFERAE: A Bestiary in Five Fingers* (Minneapolis: University of Minnesota Press, 2012).

36 For instance, Henry Buller offers an overview of ethical developments in animal studies that crystallizes common ways in which CAS is portrayed. Buller argues that the need for a *critical* animal studies is predicated on a mischaracterization of an apolitical mainstream animal studies. More significantly, he characterizes CAS as undercutting its own aims by appealing to biological distinctions between species (rather than acknowledging multispecies entanglements) and by predicating its ethics on normative, humanist models of subjectivity at precisely the moment these models need to be unsettled. These arguments are underlined with reference to Haraway's criticism of totalizing ethical frameworks. To give another example of how work in CAS can be bracketed to one side: aside from Adams, work perceived as belonging to CAS is often not engaged with substantively within so-called mainstream animal studies. More, the extensive citing of Adams carries a distinct citational politics. As Carrie Hamilton argues, attention needs to be paid to the "cu-

mulative effect of citing Adams as *the* authority on veganism and feminism," as this has resulted in other conceptually provocative work being neglected. Henry Buller, "Animal Geographies III: Ethics," *Progress in Human Geography* 40, no. 3 (2016): 425; and Carrie Hamilton, "Sex, Work, Meat: The Feminist Politics of Veganism," *Feminist Review* 114, no. 1 (2016): 112–129.

37 Again, *community* is used here in an expansive, more-than-human sense.

38 Maria Puig de la Bellacasa, "Matters of Care in Technoscience: Assembling Neglected Things," *Social Studies of Science* 41, no. 1 (2011): 91.

39 Puig de la Bellacasa, "Matters of Care," 91.

40 See, for instance, Joanna Latimer's critique of Haraway's "being with." Latimer suggests that Haraway's emphasis on relationality and hybridity can neglect the power-laden nature of particular encounters, and calls instead for preserving the "possibility of dwelling with non-humans" in ways that preserve "division and alterity as much as connectivity and unity." Joanna Latimer, "Being Alongside: Rethinking Relations amongst Different Kinds," *Theory, Culture and Society* 30, no. 7–8 (2013): 98.

41 Rosemary-Claire Collard, "Putting Animals Back Together, Taking Commodities Apart," *Annals of the Association of American Geographers* 104, no. 1 (2015): 153.

42 Franklin Ginn, "Sticky Lives: Slugs, Detachment and More-than-Human Ethics in the Garden," *Transactions of the Institute of British Geographers* 39, no. 4 (2014): 538.

43 This argument is revisited throughout the book, but I elaborate on it in particular depth in the conclusion.

44 Jo Freeman, "The Tyranny of Structurelessness," in *Untying the Knot: Feminism, Anarchism and Organisation*, by Jo Freeman and Cathy Levine (London: Dark Star and Rebel Press, 1984), 6–7.

45 I have explored the way Freeman's ideas have intersected with pedagogy in Eva Giraud, "Feminist Praxis, Critical Theory and Informal Hierarchies," *Journal of Feminist Scholarship* 7, no. 1 (2015): 43–60. See chapter 2 (in this book) for elaboration of how these arguments have been related to digital culture.

46 Freeman, "Tyranny of Structurelessness," 16.

47 Ginn, "Sticky Lives," 533.

48 Puig de la Bellacasa, "Matters of Care," 91.

49 Barad, *Meeting the Universe Halfway*, 71. As Felicity Callard and Des Fitzgerald's work elucidates, Barad's use of *entanglement* can be especially productive in the context of knowledge production: drawing attention to more expansive apparatuses through which knowledge is produced, to the relationships between different bodies of knowledge, and to affective dimensions of research. This specific use of *entanglement* contrasts, however, with the more wide-ranging uses of the term discussed throughout the book, to characterize environmental and multispecies relations (only some of which, such as Haraway, engage explicitly with Barad). Again, moreover, the issue is not with conceptions of entanglement in themselves but with particular ethical claims that have been grounded in appeals to irreducible complexity and relationality. Des Fitzgerald and Felicity Callard, "Social Science and Neuroscience beyond Interdisciplinarity: Experimental Entanglements," *Theory, Culture and*

Society 32, no. 1 (2015): 3–32; and Callard and Fitzgerald, *Rethinking Interdisciplinarity across the Social Sciences and Neurosciences*.

50 Though the work of a range of different thinkers is synthesized in developing these arguments, Haraway and Susan Leigh Star have the greatest prominence in chapter 1; Karen Barad and Anna Tsing are the focus in the second chapter; and Isabelle Stengers's cosmopolitical approach is at the heart of chapter 3.

51 Again a range of theorists and interlocutors in key debates are brought together in the final three chapters, but—broadly speaking—central to the fourth chapter is a sustained reading of Maria Puig de la Bellacasa's conception of care against work in more-than-human geographies; and the final chapters draw together work focused on nonhuman charisma (notably Jamie Lorimer) and bodily encounters (such as Vinciane Despret) in particular.

52 The value of emotion has been reiterated by a number of thinkers, particularly in relation to developing sustainable forms of activism, for example, Paul Chatterton and Jenny Pickerill, "Everyday Activism and the Transitions towards Post-capitalist Worlds," *Transactions of the Institute of British Geographers* 35, no. 4 (2010): 475–490; for further discussion of emotion and its value in engaging with publics, but also the ways it can be used to delegitimize activist perspectives, see Jeffrey Juris, "Performing Politics: Image, Embodiment, and Affective Solidarity during Anti-corporate Globalization Protests," *Ethnography* 9, no. 1 (2008): 61–97.

53 A large body of recent research has highlighted the active role of care, emotion, and affective labor (and the relations between these qualities) within experimental research, in ways that belie the rational/irrational dichotomy between researchers and activists, for example, Martyn Pickersgill, "The Co-production of Science, Ethics, and Emotion," *Science, Technology, and Human Values* 37, nos. 6 (2012): 579–603; and Des Fitzgerald, "The Affective Labour of Autism Neuroscience," *Subjectivity* 6 (2013): 131–152.

54 Haraway, *When Species Meet*. For an overview of these debates that engages with Haraway's influence, see Joanna Latimer and Mara Miele, "Naturecultures: Science, Affect and the Nonhuman," *Theory, Culture and Society* 30, nos. 7–8 (2013): 5–31.

55 An argument made in Haraway's *When Species Meet*, as noted above, but one that has a long history (as discussed in chapter 4 in this book).

56 It is this sort of routine characterization of certain practices as totalizing and essentialist (namely, those engaged in by activists) and others as situated, embodied, and responsive (in line with the approaches advocated in theoretical contexts) that I seek to unsettle throughout this book. Hamilton makes a similar argument in her sympathetic critique of Val Plumwood's work, suggesting that Plumwood's critique of veganism as a form of ontological purity "effectively reduces all veganism to its 'ontological' variety, failing to acknowledge alternative vegan traditions, including those within the food, environmental and social justice movements," a move that Hamilton argues "represents a form of asceticism and alienation from embodiment [that] forecloses discussion of the different embodied experiences of vegans/vegetarians." Hamilton, "Sex, Work, Meat," 122. For Plumwood's original critiques, see Val Plumwood, "Integrating Ethical Frameworks for Animals, Hu-

mans, and Nature: A Critical Feminist Eco-socialist Analysis," *Ethics and the Environment* 5, no.2 (2000): 285–322; and Val Plumwood, "Gender, Eco-feminism and the Environment," in *Controversies in Environmental Sociology*, edited by Richard White (Cambridge: Cambridge University Press, 2004), 43–60.

57 My primarily theoretical concerns here are the reason why I have focused on exploring the conceptual potentials of these tactics rather than elaborating on the specificities of each movement. For a sustained discussion of anticapitalist and environmental protest in the United Kingdom that provides further ethnographic context, see Chatterton and Pickerill, "Everyday Activism." For further ethnographic detail about some of the key movements discussed here, see Eva Giraud, "Displacement, 'Failure' and Friction: Tactical Interventions in the Communication Ecologies of Anti-capitalist Food Activism," in *Digital Food Activism*, ed. Tanja Schneider et al. (New York: Routledge, 2018), 130–150; and Eva Giraud, "Practice as Theory: Learning from Food Activism and Performative Protest," in *Critical Animal Geographies: Politics, Intersections and Hierarchies in a Multispecies World*, ed. Kathryn Gillespie and Rosemary-Claire Collard (New York: Routledge, 2015), 36–53.

58 Michel de Certeau, *The Practice of Everyday Life*, trans. Steven Rendall (Berkeley: University of California Press, 1984); for a use of this approach to articulate relations with more-than-human actors, see John Law and Annemarie Mol, *Complexities: Social Studies of Knowledge Practices* (Durham, NC: Duke University Press, 2002).

59 Latour, "Why Has Critique Run Out of Steam?," 229.

60 De Certeau, *Practice of Everyday Life*, 30–37.

61 Donna Haraway, "The Promises of Monsters: A Regenerative Politics for Inappropriate/d Others," in *Cultural Studies*, ed. Lawrence Grossberg, Cary Nelson, and Paula Treichler (New York: Routledge, 1992), 311.

1. *Articulations*

1 Although Star's precise articulation of this point was in 1991, this is regularly cited as one of the central concerns of feminist science studies (a point made by Maria Puig de la Bellacasa in "Matters of Care") as well as science studies more broadly. See Star, "Power," 38.

2 For an elaboration of why this process of articulation is so difficult for activists, see Pollyanna Ruiz, *Articulating Dissent: Protest and the Public Sphere* (London: Pluto, 2014).

3 As Dimitris Papadopoulos argues, seemingly mundane elements of activist practice often carry significance, as it is often at the everyday level—"beyond the radar of control"—that "creative social transformation" takes place. Correspondingly, barriers that inhibit practice at this mundane, everyday level also hold ethico-political importance, due to their capacity to *undermine* these transformative potentials. Dimitris Papadopoulos, *Experimental Practice: Technoscience, Alterontologies, and More-than-Social Movements* (Durham, NC: Duke University Press, 2018), 4.

4 While this framing of constraints as productive is Foucauldian in tone, I am drawing more explicitly on feminist-materialist rereadings of Foucault and Judith Butler that emphasize the relations between the material and the semiotic, and how these relations are productive of matter as well as discursive regimes of truth. For an

elaboration of this distinction, see Karen Barad, "Getting Real: Technoscientific Practices and the Materialization of Reality," *Differences* 10, no. 2 (1998): 87–128.

5 To reiterate a point made in the introduction, this line of argument resonates with concerns put forward by Puig de la Bellacasa in "Matters of Care."

6 For an especially helpful overview of the relationships among McSpotlight, the McInformation Network, and other anticapitalist movements, see Jenny Pickerill, *Cyberprotest: Environmental Activism Online* (Manchester: Manchester University Press, 2003).

7 Concerns about activists' increasing use of social media, such as Facebook and Twitter, rather than activist-led alternative digital media have gathered force throughout the 2000s. These tensions are discussed in depth in the next chapter, but for a broad overview of shifts from alternative online media to proprietary software platforms, see Joss Hands, Greg Elmer, and Ganaele Langlois, eds., "Platform Politics," special issue, *Culture Machine* 14 (2013), https://monoskop.org/images/c/ce/Culture_Machine_Vol_14_Platform_Politics.pdf.

8 See again Ruiz, *Articulating Dissent*.

9 Within this chapter I am working with the account of *articulation* put forward by Donna Haraway in "The Promises of Monsters." For a distinction between this conception of the term and the alternative cultural studies tradition of *articulation* found in the work of Stuart Hall, which focuses on the articulation of different elements together in the creation of hegemony (or indeed counterhegemony), see Nathan Stormer, "Articulation: A Working Paper on Rhetoric and Taxis," *Quarterly Journal of Speech* 90, no. 3 (2004): 257–284.

10 London Greenpeace was not the same as the international movement, and actually predated it.

11 Originally six activists were sued, but the others felt pressured into apologizing. Steel, however, refused, and this led to her and Morris being sued for libel. The pamphlet was subsequently pared down into an A5 leaflet entitled *What's Wrong with McDonald's?*, which continues to be distributed today. For further technical detail, see John Vidal, *McLibel: Burger Culture on Trial* (Chatham, UK: Pan Books, 1997).

12 For the full story, told in the activists' words, see "The McLibel Trial Story," *McSpotlight*, McInformation Network, accessed January 18, 2019, http://www.mcspotlight.org/case/trial/story.html.

13 Chris Atton, "Reshaping Social Movement Media for a New Millennium," *Social Movement Studies* 2, no. 1 (2003): 3–15; and Chris Atton, *Alternative Media* (London: Sage, 2002), 144–150. Debates about McSpotlight's capacity to support protest narratives are developed further in the next chapter.

14 Star, "Power," 37.

15 Star, "Power," 40.

16 This book tends to use *friction* in the broader sense in which it has been engaged with in media and cultural theory, as designating tensions associated with particular infrastructures, when these infrastructures break down, fail to work, or come into conflict with the messiness of practice. The frictions focused on in this chapter, however, can be understood in the more specific sense outlined by Anna Tsing, when she describes *friction* as referring to the tensions that arise when universal-

izing norms—themselves a product of attempts to standardize and scale up particular infrastructures in support of global capitalism—are materialized in situated local contexts. Anna Tsing, *Friction: An Ethnography of Global Connection* (Princeton, NJ: Princeton University Press, 2005).

17 Haraway, "Species Matters."

18 For a particularly informative overview of these developments in more-than-human geographies, see Lorimer, "Multinatural Geographies," or, in relation to the environmental humanities, Thom van Dooren, Eben Kirksey, and Ursula Münster, "Multispecies Studies: Cultivating Arts of Attentiveness," *Environmental Humanities* 8, no. 1 (2016): 1–23.

19 Haraway, "Promises of Monsters," 309.

20 The postcolonial commitments of Haraway's early work are vital to reassert in light of criticisms that have been articulated, particularly by Indigenous scholars as well as more broadly within contemporary feminist and queer scholarship, toward more recent conceptual approaches that have emphasized nonhuman agency but either failed to engage with or actively obscured the antiracist roots of early work in the field. These problems have been the focus of important interventions by scholars who have sought to decolonize this body of theoretical work, such as the work of Kim TallBear and Zoe Todd. A number of other scholars, including Sari Irni, Juanita Sundberg, Marie Thompson, and Angela Willey, have also argued that theoretical work such as posthumanism, new materialism, actor-network theory, and more-than-human geography can naturalize particular Western European and Anglo-American traditions of thought, even while aspiring to do the opposite. In addition to Sundberg's aforementioned "Decolonizing Posthumanist Geographies," see Sari Irni, "The Politics of Materiality: Affective Encounters in a Transdisciplinary Debate," *European Journal of Women's Studies* 20, no. 4 (2013): 347–360; TallBear, "Beyond the Life/Not Life Binary"; Todd, "Indigenous Feminist's Take"; Marie Thompson, "Whiteness and the Ontological Turn in Sound Studies," *Parallax* 23, no. 3 (2017): 266–282; and Angela Willey, "A World of Materialisms: Postcolonial Feminist Science Studies and the New Natural," *Science, Technology, and Human Values* 41, no. 6 (2016): 991–1014. For an excellent collection of articles focused on decolonizing science studies, see Anne Pollock and Banu Subramaniam, eds., "Resisting Power, Retooling Justice," special issue, *Science, Technology, and Human Values* 14, no. 6 (2016).

21 The meme also feeds into critical work about the affordances of social media, which demonstrate the limitations of an affective politics premised purely on the circulation of imagery, as discussed in more depth in the next chapter and chapter 5. See, for instance, Jodi Dean, *Democracy and Other Neoliberal Fantasies: Communicative Capitalism and Left Politics* (Durham, NC: Duke University Press, 2009).

22 For the meme and related criticisms, see Angela Vandenbroek, "This Image Should NOT Be Seen by the Whole World," *How to Be an Anthropologist* (blog), April 15, 2013, http://ak.vbroek.org/2013/04/15/this-image-should-not-be-seen-by-the-whole-world/.

23 For a discussion of how the rush to denunciate has a detrimental effect on politics, see Isabelle Stengers, *In Catastrophic Times: Resisting the Coming Barbarism*, trans., Andrew Goffey (Paris: Open Humanities Press, 2015).

24 In "Being Alongside," Joanna Latimer (drawing on Marilyn Strathern) offers an informative discussion of how particular anthropocentric and ethnocentric modes of representation are bound up with broader social "orderings of relations."
25 Haraway, "Promises of Monsters," 312. Haraway does not reproduce the problematic images she criticizes here, to avoid perpetuating their logic. Likewise, I am not reproducing the images used in the original social media post here.
26 Haraway, "Promises of Monsters," 312.
27 Lorimer, *Wildlife in the Anthropocene*, 78. I discuss in chapter 6 how these tensions are manifested in awareness-raising films.
28 Haraway, "Promises of Monsters," 309.
29 Again see Vandenbroek, "This Image."
30 As a follow-up blog post makes clear, the image was unrelated to the hydroelectric project and was instead depicting a family reunion. Angela Vandenbroek, "It Isn't Just about the Falsehood—Follow Up on the Chief Raoni Crying Meme," *How to Be an Anthropologist* (blog), April 20, 2013, http://ak.vbroek.org/2013/04/20/it-isnt-just-about-the-falsehood/.
31 The obvious difficulty in this task lies in working with actors who cannot offer responses that are intelligible within a humanist framework. Affect and care, which are often seen as arising from bodily engagements, have been framed as a means of overcoming this difficulty within animal studies and the environmental humanities. This approach frequently takes a lead from Haraway and Vinciane Despret. While these arguments will be turned to in depth in chapters 4 and 5, the focus in this chapter is on the more fundamental implications of a conceptual emphasis on relational ethics.
32 Haraway subsequently articulates arguments in favor of relational ethics by drawing on the vocabulary of intra-action and entanglement, as popularized by Karen Barad (for more on this relationship see chapter 2). See: Haraway, *When Species Meet*.
33 Susanna Hecht and Alexander Cockburn, *The Fate of the Forest: Developers, Destroyers, and Defenders of the Amazon* (Chicago: University of Chicago Press, 1990).
34 Haraway, "Promises of Monsters," 309.
35 The concept of sympoiesis is central to Haraway's *Staying with the Trouble*.
36 Haraway, "Promises of Monsters," 311.
37 William Lynn, "Animals, Ethics and Geography," in *Animal Geographies: Place, Politics, and Identity in the Nature-Culture Borderlands*, ed. Jennifer R. Wolch and Jody Emel (New York: Verso, 1998), 281.
38 On the importance of situating dualistic thought as a liberal-humanist concern, rather than applicable to thought in general (and thus inadvertently contributing to the erasure of different epistemologies), see TallBear, "Beyond the Life/Not Life Binary"; Todd, "Indigenous Feminist's Take."
39 For a discussion of anthropocentric orderings of relations that are predicated on epistemological dualisms, see again Latimer, "Being Alongside."
40 This text is from the original McDonald's fact sheet (as opposed to the more recent, abbreviated, *What's Wrong with McDonald's?* leaflet). The full text can be found here: "The Out of Print Original Fact Sheet," McSpotlight, McInformation Network, last accessed January 18, http://www.mcspotlight.org/case/factsheet.html.

41 This text is again from "The Out of Print Original Fact Sheet."

42 Philippe Pignarre and Isabelle Stengers, *Capitalist Sorcery: Breaking the Spell*, trans. Andrew Goffey (London: Palgrave Macmillan, 2011), 30. See also the section of the same book entitled "Sorry, But We Have To," 89–94.

43 Stengers, *In Catastrophic Times*, 31.

44 For details about UK libel laws, see Marlene Arnold Nicholson, "McLibel: A Case Study in English Defamation Law," *Wisconsin International Law Journal* 18 (2000): 1–144.

45 Steve Woolgar and Javier Lezaun, "The Wrong Bin Bag: A Turn to Ontology in Science and Technology Studies?," *Social Studies of Science* 43, no. 3 (2013): 332.

46 Vidal, *McLibel*, 217.

47 Woolgar and Lezaun, "Wrong Bin Bag," 333.

48 "Day 179—30 Oct 95—Page 07," transcript, McSpotlight, McInformation Network, October 30, 1995, http://www.mcspotlight.org/cgi-bin/zv/case/trial/transcripts/951030/07.htm; and "Day 191—24 Nov 95—Page 04," transcript, McSpotlight, McInformation Network, November 24, 1995, http://www.mcspotlight.org/cgi-bin/zv/case/trial/transcripts/951124/04.htm.

49 Vidal's *McLibel* has an extensive discussion of the verdict, which emphasizes this point. Vidal, *McLibel*, 312.

50 The full transcript of the verdict is available at "The Verdict," McSpotlight, McInformation Network, June 19, 1997, http://www.mcspotlight.org/cgi-bin/zv/case/trial/verdict/verdicto_sum.html.

51 David Wolfson, *The McLibel Case and Animal Rights* (London: Active Distribution, 1999), 20.

52 Wolfson, *McLibel Case*, 19.

53 Wolfson, *McLibel Case*, 19.

54 Vidal, *McLibel*, 311.

55 As I touch on in note 31, a burgeoning body of work has addressed this problem through a focus on somatic, bodily encounters as a means of articulating with nonhuman partners. Though this line of inquiry has proven fruitful, a number of thinkers have raised the question of its limitations in contexts where the nonhuman actors at stake are inaccessible, due to occurring in spaces separated (or even hidden) from public view. Despite their own commitment to this approach, for instance, Beth Greenhough and Emma Roe note that a focus on the microsociological, somatic level—though productive—does not address the question of "how these kinds of sensibility might operate at a distance"; they also warn that "emphasising proximate and immediate relations also obscures historical and temporal dimensions" of certain uses of animals (their example being animal research) "and the underlying power relations they inscribe." These debates are explored in further depth in chapters 4 and 5 in particular. See Beth Greenhough and Emma Roe, "Ethics, Space, and Somatic Sensibilities: Comparing Relationships between Scientific Researchers and Their Human and Animal Experimental Subjects," *Environment and Planning D: Society and Space* 29, no. 1 (2011): 47–66; and Vinciane Despret, "Responding Bodies and Partial Affinities in Human-Animal Worlds," *Theory, Culture and Society* 30, nos. 7–8 (2013): 51–76.

56 Puig de la Bellacasa, "Matters of Care," 92–94; the example focused on in this article, however, is not unskilled restaurant labor but domestic and care work.

57 See Nicole Shukin, *Animal Capital: Rendering Life in Biopolitical Times* (Minneapolis: University of Minnesota Press, 2009); or Chris Philo, "Animals, Geography, and the City: Notes on Inclusions and Exclusions," *Environment and Planning D: Society and Space* 13, no. 6 (1995): 655–681.

58 Star, "Power," 43.

59 Star, "Power," 48.

60 The section of McSpotlight entitled "Employment" includes a large number of critical press reports about McDonald's working conditions, including an article by Danny Penman, "McDonald's Staff 'Worked among Sewage,'" *Independent*, October 31, 1995, http://www.mcspotlight.org/media/press/sewage.html. Available from "Employment," McSpotlight, McInformation Network, last accessed January 18, 2019, http://www.mcspotlight.org/issues/employment/index.html.

61 Vidal, *McLibel*, 115.

62 These frictions again resonate with Tsing's arguments in *Friction*.

63 For an overview of work that has argued for the politics of foregrounding friction, see Giraud, "Displacement, 'Failure' and Friction."

64 Alaimo, *Exposed*, 137.

65 These potentials for dialogue between strands of theory and practice are fleshed out in a more sustained way in Papadopoulos's *Experimental Practice*.

66 I take the term *ontological politics* here from Annemarie Mol, "Ontological Politics: A Word and Some Questions," in *Actor Network Theory and After*, ed. John Law and John Hassard (Oxford: Blackwell, 1999), 74–89.

67 The lack of legal aid was one of the key (though not the primary) reasons why the case was seen as such a David-and-Goliath battle. In the United Kingdom at the time, defendants were denied legal aid in libel cases, which is why Steel and Morris had to represent themselves, as they could not afford to pay for legal representation. The disparity between their financial plight and the resources available to McDonald's is discussed in depth in Vidal's *McLibel*.

68 See "McLibel: Longest Case in English History," BBC *News*, February 15, 2005, http://news.bbc.co.uk/2/hi/uk_news/4266741.stm.

69 All of the witness statements can be found here: "The McLibel Trial: Evidence," McSpotlight, McInformation Network, accessed May 5, 2018, http://www.mcspotlight.org/case/trial/evidence.html.

70 Franny Armstrong and Ken Loach, dirs., *McLibel: Two People Who Wouldn't Say Sorry* (London: Spanner Films, 2005).

71 Hecht's witness statement can be found archived on the McSpotlight website here: "Witness Statement," McSpotlight, McInformation Network, February 21, 1996, http://www.mcspotlight.org/people/witnesses/environment/hecht_susanna.html.

72 This is not to say Hecht's nuanced approach to articulating issues surrounding deforestation is intrinsically problematic; indeed, quite the opposite: her testimony points to a problem in the legal system in its insistence on certain forms of evidence. It is still the case, however, that the requirements of sociolegal environ-

ments place restrictions on what can be said and who can say it, which cannot be dismissed in theoretical contexts.

73 Cary Wolfe, *What Is Posthumanism?* (Minneapolis: University of Minnesota Press, 2009).

74 Witness statement, available from "Witness Statement: Kenneth Miles," McSpotlight, McInformation Network, January 11, 1994, http://www.mcspotlight.org/people/witnesses/advertising/miles.html.

75 Mol, "Ontological Politics," 79.

76 The role of exclusion in ontological politics is emphasized by Richie Nimmo, *Actor Network Theory Research* (London: Sage, 2016), xxxii.

77 Puig de la Bellacasa, "Matters of Care," 91.

78 "Case of Steel and Morris v. the United Kingdom," European Court of Human Rights, May 15, 2005, http://hudoc.echr.coe.int/eng?i=001-68224.

79 Henry Buller, "Animal Geographies I," *Progress in Human Geography* 38, no. 2 (2014): 314.

80 Shotwell, *Against Purity*, 117.

81 Haraway, *When Species Meet*, 3.

82 Ginn, "Sticky Lives"; and Freeman, "Tyranny of Structurelessness."

2. *Uneven Burdens of Risk*

1 Pollyanna Ruiz's work again provides rich insight into the relationship between media technologies and protest movements, in relation to this question of articulating connections among issues. See Ruiz, *Articulating Dissent*.

2 CounterSpin Collective, "Media, Movement(s) and Public Image(s): Counterspinning in Scotland," in *Shut Them Down!: The G8, Gleneagles 2005 and the Movement of Movements*, ed. David Harvie et al. (Leeds, UK: Dissent!; Brooklyn, NY: Autonomedia, 2005), 322–323.

3 "Look like" is in quotation marks here as it is a refrain taken from Rodrigo Nunes's reflections on horizontal activism in the wake of the 2005 anti-G8 protests. His point in the article resonates with the arguments outlined here in pointing to the danger of rigid models and practices congealing as what democracy is thought to "look like." Rodrigo Nunes, "Nothing Is What Democracy Looks Like," in Harvie et al., *Shut Them Down!*, 299–320.

4 For example, Joss Hands, "Civil Society, Cosmopolitics and the Net: The Legacy of 15 February 2003," *Information, Communication and Society* 9, no. 2 (2006): 225–243.

5 Anna Tsing, *The Mushroom at the End of the World: On the Possibility of Life in Capitalist Ruins* (Princeton, NJ: Princeton University Press, 2015), 40–41.

6 Tsing, *Mushroom*, 40–41.

7 Star, "Power."

8 George Ritzer, *The McDonaldization of Society* (London: Sage, 2013).

9 For arguments in favor of an ethics of storytelling, see van Dooren, *Flight Ways*; and Haraway, *Staying with the Trouble*.

10 Haraway, *Staying with the Trouble*, 97.

11 Isabelle Stengers, "The Cosmopolitical Proposal," in *Making Things Public: Atmo-*

spheres of Democracy, ed. Bruno Latour and Peter Weibel (Cambridge, MA: MIT Press, 2005), 996.

12 Chatterton and Pickerill, "Everyday Activism,"485.

13 Chatterton and Pickerill, "Everyday Activism," 485–486.

14 For explicit discussion of deterritorialization, see Jonathan Bach and David Stark, "Link, Search, Interact: The Co-evolution of NGOs and Interactive Technology," *Theory, Culture and Society* 21, no. 3 (2004): 101–117. Work by Manuel Castells was a touchstone in more positive appraisals of the internet in the 1990s (though he has since been subject to staunch criticism); for example, Manuel Castells, *The Power of Identity* (Malden, MA: Blackwell, 1997). For a more nuanced social movement studies perspective, see Jeffrey Juris, *Networking Futures: The Movements against Corporate Globalization* (Durham, NC: Duke University Press, 2007).

15 Naomi Klein, *No Logo* (London: Flamingo, 2000), 365–396.

16 For an overview see Richard Kahn and Doug Kellner, "New Media and Internet Activism: From the 'Battle of Seattle' to Blogging," *New Media and Society* 6, no. 1 (2004): 87–95. Other key texts referred to throughout this chapter, such as Pickerill's and Juris's work, provide further context from a social movement studies perspective. Similarly, work referred to in the previous chapter, such as Philippe Pignarre and Isabelle Stengers's *Capitalist Sorcery*, refers to the Battle of Seattle as a privileged moment in conceptual as well as political terms.

17 For a series of arguments in relation to the capacity of digital media to support protest, see Joss Hands, *@ Is for Activism: Dissent, Resistance and Rebellion in a Digital Culture* (London: Pluto, 2010).

18 Nunes, "Nothing," 300.

19 CounterSpin Collective, "Media," 321 (emphasis added).

20 John Downey and Natalie Fenton, "New Media, Counter-publicity and the Public Sphere," *New Media and Society* 5, no. 2 (2003): 196.

21 For an analysis of the difficulties of engaging in polyvocal protest, which brings together diverse voices, see Ruiz, *Articulating Dissent*.

22 This language of utopia, for instance, is employed in Pip Shea, Tanya Notley, and Jean Burgess's otherwise very helpful exploration of frictions between activists and media technologies in their editorial in a special issue of *Fibreculture*. Pip Shea, Tanya Notley, and Jean Burgess, "Editorial: Entanglements—Activism and Technology," *Fibreculture* 26 (2015): 1–6.

23 Pickerill, *Cyberprotest*, 31.

24 Pickerill, *Cyberprotest*, 31.

25 Pickerill, *Cyberprotest*, 49.

26 Pickerill, *Cyberprotest*, 36.

27 As described in the previous chapter, Stengers is intensely critical of the logic of "sorry, but we have to" within *In Catastrophic Times*.

28 Hilde Stephansen and Emiliano Treré, "From 'Audiences' to 'Publics': The Value of a Practice Framework for Research on Alternative and Social Movement Media" (paper presented at European Communication Research and Education Association [ECREA], Charles University, Prague, November 10, 2016).

29 Pickerill discusses this subsequently in relation to Indymedia Australia, in "'Au-

tonomy Online': Indymedia and Practices of Alter-Globalisation," *Environment and Planning A: Economy and Space* 39, no. 11 (2007): 2668–2684.

30 Pickerill, *Cyberprotest*, 10.

31 Veronica Barassi, "Ethnographic Cartographies: Social Movements, Alternative Media and the Space of Networks," *Social Movement Studies* 12, no. 1 (2013): 58.

32 I tease out the significance of these incidents, and of the ways activists navigate tensions that arise from them, in Giraud, "Displacement, 'Failure' and Friction."

33 A claim made by activists; see *What's Wrong with McDonald's?*, fact sheet, McSpotlight, 1986, http://www.mcspotlight.org/case/factsheet.html.

34 Indymedia reports on McDonald's protests, many of which take the form of food giveaways (a practice discussed in more depth in the next chapter), can be found by performing a site-specific search in Google, using "free food nottingham site:indymedia.org.uk": https://www.google.co.uk/search?as_sitesearch=indymedia.org.uk&as_q=free+food+nottingham&gws_rd=ssl.

35 For further elaboration of this argument, see again Giraud, "Displacement, 'Failure' and Friction." For discussion of the mundane "remix" aesthetics of activist websites: Adam Fish, "Mirroring the Videos of Anonymous: Cloud Activism, Living Networks, and Political Mimesis," *Fibreculture* 26 (2015): 85–107.

36 For example, Marc Garcelon, "The 'Indymedia' Experiment: The Internet as Movement Facilitator against Institutional Control," *Convergence: The International Journal of Research into New Media Technologies* 12, no. 1 (2006): 55–82. Indymedia is also described in terms of an experimental laboratory in Pickerill, "'Autonomy Online.'"

37 For instance, Indymedia was regularly a key case study of the impact of digital media on news production in mainstream media studies textbooks; for example, Stuart Allan, *Online News: Journalism and the Internet* (Maidenhead, UK: Open University Press, 2006).

38 A comprehensive overview of Indymedia's history can be found in Victor Pickard, "Assessing the Radical Democracy of Indymedia: Discursive, Technical, and Institutional Constructions," *Critical Studies in Media Communication* 23, no. 1 (2006): 19–38.

39 Kahn and Kellner, "New Media."

40 Although these links with the global justice movement were generally the case, the network was designed to extend beyond this narrow membership, and local centers did gain distinct identities. For a discussion of some of the variations within the different groups involved, see Garcelon, "'Indymedia' Experiment."

41 Nico Carpentier, Peter Dahlgren, and Francesca Pasquali, "Waves of Media Democratization: A Brief History of Contemporary Participatory Practices in the Media Sphere," *Convergence: The International Journal of Research into New Media Technologies* 19, no. 3 (2013): 287–294.

42 Pickerill, "'Autonomy Online.'"

43 Ippolita, Geert Lovink, and Ned Rossiter, "The Digital Given: 10 Web 2.0 Theses," *Fibreculture* 14 (2009), accessed February 1, 2019, http://fourteen.fibreculturejournal.org/fcj-096-the-digital-given-10-web-2-0-theses/.

44 Leah Lievrouw, *Alternative and Activist New Media* (London: Polity, 2011).

45 For tables that detail these findings in full, see Eva Giraud, "Has Radical Partici-

patory Online Media Really 'Failed'? Indymedia and Its Legacies," *Convergence: The International Journal of Research into New Media Technologies* 20, no. 4 (2014): 425.

46 Todd Wolfson, "From the Zapatistas to Indymedia: Dialectics and Orthodoxy in Contemporary Social Movements," *Communication, Culture and Critique* 5 (2012): 149–170.

47 Freeman, "Tyranny of Structurelessness," 6–7.

48 A discussion of attempts at outreach can be found in Pickerill's aforementioned article about Indymedia in Australia, "'Autonomy Online.'" I synthesize some of the difficulties in providing sustained outreach plans, as articulated by a range of social movement theorists, in my later analysis of difficulties facing Indymedia in "Has Radical Participatory Online Media Really 'Failed'?"

49 T. Wolfson, "From the Zapatistas."

50 For a collection of essays that gives a helpful overview of the early years after the 1994 uprising, see John Holloway and Elena Peláez, *Zapatista! Reinventing Revolution in Mexico* (London: Pluto, 1997); for a discussion of the international dimension of the campaign and its early legacies, see Thomas Oleson, *International Zapatismo: The Construction of Solidarity in the Age of Globalization* (London: Zed Books, 2005).

51 Joss Hands describes some of the legacies of Zapatista uses of the internet in *@ Is for Activism*, and a prominent early example can be found in Castells's *The Power of Identity*. For one of the earliest commentaries, see Harry Cleaver, "The Chiapas Uprising and the Future of Class Struggle in the New World Order," February 14, 1994, http://www.eco.utexas.edu/facstaff/Cleaver/chiapasuprising.html.

52 Haraway, *Staying with the Trouble*, 155.

53 As underlined in Juanita Sundberg's engagements with Zapatista philosophy in "Decolonizing Posthumanist Geographies," this decontextualization is antithetical to the Zapatistas' work.

54 For a collection of Zapatista communiqués, see Ziga Vodovnik, ed., *¡Ya Basta! Ten Years of the Zapatista Uprising* (Oakland, CA: AK Press, 2004).

55 T. Wolfson, "From the Zapatistas," 153.

56 T. Wolfson, "From the Zapatistas," 164.

57 Fabian Frenzel et al., "Comparing Alternative Media in North and South," *Environment and Planning A: Economy and Space* 43, no. 5 (2012): 1173–1189.

58 Frenzel et al., "Comparing Alternative Media," 1183.

59 Frenzel et al., "Comparing Alternative Media," 1183.

60 Giraud, "Has Radical Participatory Online Media Really 'Failed'?," 427.

61 Garcelon, "'Indymedia' Experiment."

62 Scott Uzelman, "Media Commons and the Sad Decline of Vancouver Indymedia," *Communication Review* 14, no. 4 (2011): 285.

63 Uzelman, "Media Commons," 286.

64 For an overview of these issues, see Giraud, "Has Radical Participatory Online Media Really 'Failed'?"

65 Shea, Notley, and Burgess, "Editorial."

66 What Barad explicitly draws attention to are the moments in which discrete identities emerge through specific assemblages. These agential cuts are elucidated by a

range of different examples in Barad's work; for instance, at the atomic level particular laboratory apparatuses can enact a "cut" wherein light is enacted as either a particle or a wave, depending on the apparatus through which it is measured. A more everyday example is offered by Barad's description of exploring a dark room with a stick; depending on how the room, stick, and navigator relate to one another, the stick could be part of the measuring apparatus (used to navigate the room) or part of the room being negotiated (an entity that is itself felt and measured by someone grappling in the darkness). We elaborate on the ethical provocations offered by these cuts in Hollin et al., "(Dis)entangling Barad."

67 Anna Feigenbaum, "Resistant Matters: Tear Gas, Tents and the 'Other Media' of Occupy," *Communication and Critical/Cultural Studies* 11, no. 1 (2014): 21.

68 Relational understandings of media technologies are also central to media-ecological perspectives, an approach that Feigenbaum draws on.

69 Again, this is helpfully encapsulated by Feigenbaum's work.

70 Barad's emphasis on the stability of agential cuts once they have occurred, for instance, gives her approach a slightly different emphasis than Mol's focus on the reality effects emerging from particular sociotechnical performances or Zylinska's relationality without cuts. See Annemarie Mol, *The Body Multiple: Ontology in Medical Practice* (Durham, NC: Duke University Press, 2002); and Joanna Zylinska, *Minimal Ethics for the Anthropocene* (Ann Arbor, MI: Open Humanities Press, 2014).

71 For a reflection on Indymedia's legacies, see Tish Stringer, "This Is What Democracy Looked Like," in *Insurgent Encounters: Transnational Activism, Ethnography, and the Political*, ed. Jeffrey Juris and Alex Khasnabish (Durham, NC: Duke University Press, 2013), 318–341.

72 For an important exploration of an ethics of experimentation in relation to knowledge production, see Des Fitzgerald and Felicity Callard, "Social Science and Neuroscience beyond Interdisciplinarity: Experimental Entanglements," *Theory, Culture and Society* 32, no. 1 (2015): 3–32. On the value of experimentation in activism, Papadopoulos, *Experimental Practice*.

73 Haraway, *When Species Meet*, 3.

74 For a discussion of debates over preferred terminology, see Alex Trocchi, Giles Redwolf, and Petrus Alamire, "Reinventing Dissent! An Unabridged Story of Resistance," in Harvie et al., *Shut Them Down!*, 61–100.

3. *Performing Responsibility*

1 Haraway, *When Species Meet*, 70–73.

2 *Beasts of Burden* (London: Active Distribution, 2004).

3 I use the term *responsibility* here as a broad umbrella encompassing the range of different ways that it and similar terms are used in reference to the techniques and practices engaged in by social movements that are designed to foster responsibility and accountability for their actions and particular ethico-political decisions. These processes include, for instance, the practices that support the "consensus, transparency, and inclusiveness" that Donatella Della Porta suggests characterize the deliberative democracy associated with the global justice movement, values

that are designed to prevent decisions from being made for others and to instead provide platforms for collective decision-making. Responsibility is also, as Jenny Pickerill and Paul Chatterton point out, fostered on a more everyday level in social movements, when activists have to negotiate between pragmatic necessities and "hoped-for" worlds. I am using *responsibility practices* in a broad sense, therefore, to encapsulate both formal participatory practices (such as consensus decision-making) and microsociological practices that foster responsibility on a more everyday basis. See Donatella Della Porta, "Making the Polis: Social Forums and Democracy in the Global Justice Movement," *Mobilization: An International Quarterly* 10, no. 1 (2005): 73–94; and Jenny Pickerill and Paul Chatterton, "Notes towards Autonomous Geographies: Creation, Resistance and Self-Management as Survival Tactics," *Progress in Human Geography* 30, no. 6 (2006): 737.

4 Félix Guattari, *The Three Ecologies*, trans. Ian Pindar and Paul Sutton (London: Continuum, 2008).

5 On cosmopolitics, see, for instance, Isabelle Stengers, "Experimenting with Refrains: Subjectivity and the Challenge of Escaping Modern Dualism," *Subjectivity* 22, no. 1 (2008): 38–59. Jamie Lorimer also advocates a cosmopolitical approach to conservation practice. Lorimer, *Wildlife in the Anthropocene*. On the turn to practice in media studies, see, for example, Nick Couldry and Andreas Hepp, *The Mediated Construction of Reality* (London: Polity, 2016).

6 Despite sharing an engagement with Guattari, for instance, the direction taken in Anna Feigenbaum, Fabien Frenzel, and Patrick McCurdy's analysis of protest camps is very different from Matthew Fuller and Andrew Goffey's analysis of "gray media" (such as spreadsheets). Anna Feigenbaum, Fabien Frenzel, and Patrick McCurdy, *Protest Camps* (London: Zed, 2013); and Matthew Fuller and Andrew Goffey, *Evil Media* (Cambridge, MA: MIT Press, 2012).

7 For descriptions of protest or media ecologies, see Feigenbaum, Frenzel, and McCurdy, *Protest Camps*; and Dan Mercea, Laura Iannelli, and Brian D. Loader, "Protest Communication Ecologies," *Information, Communication and Society* 19, no. 3 (2016): 279–289.

8 For a helpful drawing together of critical work within feminist science studies, see Sandra Harding, *The Postcolonial Science and Technology Studies Reader* (Durham, NC: Duke University Press, 2011). For overviews from the perspective of more-than-human geographies, see Lorimer, "Multinatural Geographies"; and Noel Castree et al., "Mapping Posthumanism: An Exchange," *Environment and Planning A: Economy and Space* 36, no. 8 (2004): 1341–1363. On cultural theory, see, for instance, Braidotti, *Posthuman*.

9 Van Dooren, *Flight Ways*, 45–62; and Lorimer, *Wildlife in the Anthropocene*, 159–178.

10 This argument is integral to Haraway's *When Species Meet* and *Staying with the Trouble*; see also Vinciane Despret, *What Would Animals Say If We Asked the Right Questions?* (Minneapolis: University of Minnesota Press, 2016).

11 For a sustained discussion and illustration of "conservation as cosmopolitics," see Lorimer, *Wildlife in the Anthropocene*, 97–118.

12 Steve Hinchliffe, Matthew B. Kearnes, Monica Degan, and Sarah Whatmore, "Urban Wild Things: A Cosmopolitical Experiment," *Environment and Planning D: Society and Space* 23, no. 5 (2005): 643–658.

13 Stengers, "Cosmopolitical Proposal."
14 This is the central point Haraway has made in *Staying with the Trouble*.
15 Haraway, *Staying with the Trouble*, 12.
16 Criticism of abstraction has been a central theme of Stengers's and Haraway's work, with these arguments proving influential across the environmental humanities and more-than-human geographies; Lorimer's *Wildlife in the Anthropocene* again provides an excellent example of this in fleshing out distinctions between biopolitical and cosmopolitical modes of conservation.
17 For discussions of tensions surrounding cosmopolitics and climate change, see Daniel Demeritt, "Science Studies, Climate Change and the Prospects for Constructivist Critique," *Economy and Society* 35, no. 3 (2006): 453–479. For a more sustained application of a cosmopolitical approach to climate change politics, see Stengers, *In Catastrophic Times*. With regard to animal agriculture, a key source of contention emerged in relation to Haraway's discussion of veganism versus consumption of hunted meat in *When Species Meet* (285–302), when she takes a stand for "nourishing indigestion" through refusing to take solace in vegan praxis. A number of theorists have used this as a touchstone for how perspectives such as veganism are misrepresented by relational approaches to ethics (as outlined in more depth below). Haraway has since responded to some of these criticisms in a recent interview with Sarah Franklin, acknowledging that her characterization of vegan practice may not have done justice to its messiness; see Sarah Franklin, "Staying with the Manifesto: An Interview with Donna Haraway," *Theory, Culture and Society* 34, no. 4 (2017): 49–63.
18 Weisberg, "Broken Promises of Monsters."
19 For a typical dismissal of CAS on the basis that it undercuts its own aims, see again Buller, "Animal Geographies III," 425.
20 In Stengers's terms, therefore, this stance would be seen as decisively uncosmopolitical.
21 Pedersen, "Release the Moths."
22 Lorimer, *Wildlife in the Anthropocene*, 117.
23 Though Haraway notes the possibility for care to be instrumentalized, this recognition does not fundamentally alter her approach (for example, *Staying with the Trouble*, 19). The point of this chapter is not to simply engage in a sympathetic critique of Haraway's work, however, as such engagements with noninnocence are not by any means limited to her. A recent special issue of *Social Studies of Science* that is focused on relational care ethics, for instance, discusses care's more instrumental side but, despite all of the productive reflections in the articles, does not push on the *consequences* of this noninnocence in terms of what its fundamental implications are for care ethics. More generally in the literature, and perhaps more worryingly, the acknowledgment of problems with relational ethics is usually coupled with calls to "stay with the trouble," with the result that this slogan is inadvertently positioned as an end in itself. These issues will be taken up in depth in chapter 5. Martin and colleagues' introduction to the special issue of *Social Studies of Science* provides a helpful overview of some of these issues: Aryn Martin, Natasha Myers, and Ana Viseu, "The Politics of Care in Technoscience," *Social Studies of Science* 45, no. 5 (2015): 625–641.

24 Questions about the risk of entering into particular relations will be explored in further depth in chapter 5. To summarize some of these issues, as Gregory Hollin and I foreground in our analysis of the historical consolidation of beagles as the standard laboratory dog, although all of the relationships in the first beagle colonies offered new affinities and risks for the parties involved, these risks were deeply uneven and ranged from the risk of producing bad data, to the risk of unemployment (on the part of caretakers), to death on the part of dogs who resisted entering into convivial relations with researchers. The asymmetry of the burden of these risks needs to be carefully considered in order to avoid perpetuating preexisting inequalities while giving the illusion of doing otherwise. Eva Giraud and Gregory Hollin, "Care, Laboratory Beagles and Affective Utopia," *Theory, Culture and Society* 33, no. 4 (2016): 27–49.

25 Vidal, *McLibel*, 264–294.

26 Donatella Della Porta, ed., *Global Justice Movement: Cross-national and Transnational Perspectives* (London: Routledge, 2015).

27 Stengers uses the Battle of Seattle as a key touchstone within *In Catastrophic Times*. As discussed in the previous chapter, Indymedia also emerged from these protests.

28 An excellent overview of protest camp initiatives arising from the global justice movement can be found in Feigenbaum, Frenzel, and Patrick McCurdy's *Protest Camps*.

29 A helpful collection of essays about the work undertaken at Horizone (as well as an overview of other protest activities associated with the 2005 G8) can be found in Harvie et al., *Shut Them Down!*

30 The Autonomous Geographies Collective published substantial work about autonomous activism during this period. An overview can again be found in Chatterton and Pickerill, "Everyday Activism."

31 Chatterton and Pickerill, "Everyday Activism."

32 See again Pickerill, *Cyberprotest*.

33 Haraway, for instance, begins *When Species Meet* by praising alter-globalization activists (3). As described in more depth below, in *Capitalist Sorcery*, Philippe Pignarre and Stengers see the Battle of Seattle as a moment that offers important theoretical provocations.

34 As discussed in chapter 2, Anna Feigenbaum's work offers a nuanced example of how understandings of ecology and entanglement are relevant to activist practice. Feigenbaum, Frenzel, and McCurdy's *Protest Camps* also explicitly uses this approach. In turn, a helpful account of entanglements between infrastructures and activism can be found in several articles in the special issue of *Fibreculture* edited by Pip Shea, Tanya Notley, Jean Burgess, and Su Ballard: "Entanglements—Activism and Technology," special issue, *Fibreculture* 26 (2015). The application of ecological approaches to activist media was popularized by social movement media theorists such as Emiliano Treré. Mercea et al.'s special issue of *Information, Communication and Society* also provides a helpful overview of key debates in this area. See Emiliano Treré, "Social Movements as Information Ecologies," *International Journal of Communication* 6 (2012): 2359–2377; and Dan Mercea, Laura Iannelli, and Brian D. Loader, eds., "Protest Communication Ecologies," special issue, *Information, Communication and Society* 19, no. 3 (2016).

35 Feigenbaum, Frenzel, and McCurdy, *Protest Camps*, 21.
36 Veronica Barassi's *Activism on the Web: Everyday Struggles Against Digital Capitalism* (New York: Routledge, 2015) offers a good example of this, as do a number of the different approaches outlined in the previous chapter.
37 Eloise Harding provides an excellent overview of some of the key features of these movements. Eloise Harding, "Conceptualising Horizontal Politics" (PhD diss., University of Nottingham, 2012).
38 Nunes, "Nothing," 310.
39 Haraway, *Staying with the Trouble*, 4.
40 Joan Haran also writes extensively about Starhawk's work, conceptualizing it as "imaginactivism." See Joan Haran, "Instantiating Imaginactivism: Le Guin's The Dispossessed as Inspiration," ADA: *A Journal of Gender, New Media & Technology* 12 (2017), accessed February 2, 2019, https://adanewmedia.org/2017/10/issue12-haran/.
41 Stengers, "Experimenting with Refrains," 56. The role of rituals associated with permaculture is also a particular focus of Puig de la Bellacasa's *Matters of Care: Speculative Ethics in More-than-Human Worlds* (Minneapolis: University of Minnesota Press, 2017).
42 Paul N. Edwards, "Infrastructure and Modernity: Force, Time, and Social Organization in the History of Sociotechnical Systems," in *Modernity and Technology*, ed. Thomas J. Misa, Philip Brey, and Andrew Feenberg (Cambridge, MA: MIT Press, 2003), 185–226.
43 Starhawk, "Diary of a Compost Toilet Queen," in Harvie et al., *Shut Them Down!*, 185–186.
44 Starhawk, "Diary," 194.
45 Nunes, "Nothing," 308.
46 Activists were accused of property destruction in towns such as Bannockburn and Stirling. As described by a local activist, this was leveraged to foment discontent between activists and local residents, which made it especially difficult for sympathetic members of the latter group. For local newspaper coverage of these events, see "Rampage before Dawn Ignites Protest Chaos," *Scotsman*, July 7, 2005, https://www.scotsman.com/news/rampage-before-dawn-ignites-protest-chaos-1-719084. Activist reflections can be found in Sarah, "G8 on Our Doorstep," in Harvie et al., *Shut Them Down!*, 103–108.
47 A point made, perhaps most famously, in Frantz Fanon's *Wretched of the Earth*, trans. Constance Farrington (St Ives, UK: Penguin, 1967). These debates also have long histories in anarchist praxis (quite aside from debates as to whether direct action against property even constitutes violence); as David Graeber puts it, "some will claim confrontational tactics deprive activists of the moral high ground; others will accuse those people of being elitist, and insist that the violence of the system is so overwhelming that to refuse to confront it effectively is itself acquiescence to violence." David Graeber, *Direct Action: An Ethnography* (Edinburgh: AK Press, 2009), 224.
48 *Beasts of Burden*, 29.
49 Both collectives had long-standing roles as "campaign caterers"; they not only provided food at protests but worked to produce this food in ways that prefigured non-

hierarchical anticapitalist principles (though compromises were often enacted due to particular constraints—as discussed in the rest of the chapter).

50 *Animal Liberation: Devastate to Liberate, or Devastatingly Liberal?* was originally published as an anonymous pamphlet but has since been published online at The Anarchist Library, accessed May 22, 2018, https://theanarchistlibrary.org/library/anonymous-animal-liberation-devastate-to-liberate-or-devastatingly-liberal.

51 Laura Portwood-Stacer, "Anti-consumption as Tactical Resistance: Anarchists, Subculture, and Activist Strategy," *Journal of Consumer Culture* 12, no. 1 (2012): 87–105.

52 As touched on previously, the claim that activists take moral solace in particular ethical stances (such as veganism) is a recurring refrain within *When Species Meet*, especially the final chapter, "Parting Bites," 285–302. See also note 17, this chapter.

53 Brian Dominick, *Animal Liberation and Social Revolution: A Vegan Perspective on Anarchism or an Anarchist Perspective on Veganism*, 2nd ed. (London: Active Distribution, 2008). For academic work in this area, see Stephen F. Eisenman, "The Real 'Swinish Multitude,'" *Critical Inquiry* 42, no. 2 (2016): 339–373; and Shukin, *Animal Capital*.

54 For a discussion of how racist labor relations can become entangled with the treatment of nonhuman animals, see Charlie LeDuff, "At the Slaughterhouse Some Things Never Die," in *Zoontologies: The Question of the Animal*, ed. Cary Wolfe (Minneapolis: University of Minnesota Press, 2003), 183–198.

55 A characterization of animal activism as extending rights to "privileged others" can be found in Lorimer, "Multinatural Geographies," 604.

56 See, for instance, Breeze A. Harper, *Sistah Vegan: Food, Identity, Health, and Society; Black Female Vegans Speak* (Brooklyn, NY: Lantern Books, 2010); and Sunaura Taylor, *Beasts of Burden: Animal and Disability Liberation* (New York: New Press, 2017).

57 This review of *Beasts of Burden* was originally from *Undercurrent Magazine* 8 and has since been reproduced online: Anonymous, "Beasts of Burden—Review," Libcom.org, last accessed January 22, 2019, https://libcom.org/library/beasts-of-burden-undercurrent-8.

58 Anarchist Teapot, *Feeding the Masses* (n.p.: Active Distribution, n.d.), 63–64.

59 Anarchist Teapot, *Feeding the Masses*, 62.

60 Anarchist Teapot, *Feeding the Masses*, 65.

61 Anarchist Teapot, *Feeding the Masses*, 64.

62 Anarchist Teapot, *Feeding the Masses*, 65.

63 As discussed in the previous chapter, Jo Freeman coined the term *informal hierarchy* to characterize the inadvertent hierarchies that emerge in structureless groups that have deliberately eschewed more formal hierarchies (these inadvertent hierarchies arise due to factors such as personality, friendships, or preexisting expertise). Freeman, "Tyranny of Structurelessness."

64 I provide an overview of these debates in a book chapter that focuses on the tensions associated with these protests. Giraud, "Practice as Theory."

65 This argument was made most famously by Haraway in her call to "nourish indigestion," within *When Species Meet*'s "Parting Bites" chapter, through being open to different consumption practices, rather than allowing vegan-feminist practice to become normative, but this point has since been used as a touchstone for

broader criticisms of CAS and critical-activist perspectives. (See notes 17 and 52, this chapter.)

66 I provide an overview of some of the debates about vegan feminism in Eva Giraud, "Veganism as Affirmative Biopolitics," *PhaenEx* 8, no. 2 (2014): 47–79.

67 Veggies, an activist catering collective, produced a pamphlet at a similar time to London Greenpeace and were threatened with legal action but were allowed to continue to distribute it after some minor word changes. This is discussed by John Vidal in *McLibel*, 68.

68 For a witness statement by one of Veggies' members, see "Patrick Smith Witness Statement," McSpotlight, McInformation Network, October 21, 1993, http://www.mcspotlight.org/people/witnesses/publication/veggies.html.

69 Pat Smith, "Burgers Not Bombs," *Peace News* 2514 (2009), currently available online at Peace News, accessed January 22, 2019, https://peacenews.info/node/3870/burgers-not-bombs.

70 Anarchist Teapot, *Feeding the Masses*.

71 Morgan Spurlock, dir., *Super Size Me* (Los Angeles, CA: Roadside Attractions, 2004).

72 "Vegan Free Food Giveaways," Veggies, accessed May 23, 2018, http://www.veggies.org.uk/campaigns/vegan/free-food-givaways/.

73 Keith McHenry, *Hungry for Peace: How You Can Help End Poverty and War with Food Not Bombs* (Tucson, AZ: See Sharp, 2012).

74 Nik Heynen, "Cooking Up Non-violent Civil-Disobedient Direct Action for the Hungry," *Urban Studies* 47, no. 6 (2010): 1225–1240.

75 Joshua Sbicca, "The Need to Feed: Urban Metabolic Struggles of Actually Existing Radical Projects," *Critical Sociology* 40, no. 6 (2013): 817–834.

76 Sbicca, "Need to Feed."

77 Don Mitchell and Nik Heynen, "The Geography of Survival and the Right to the City," *Urban Geography* 30, no. 6 (2009): 616.

78 Drew Robert Winter, "Doing Liberation: The Story and Strategy of Food Not Bombs," in *Anarchism and Animal Liberation*, ed. Anthony J. Nocella II, Richard J. White, and Erika Cudworth (Jefferson, NC: McFarland, 2015), 61.

79 For my original Indymedia reports of the protests, see "Vegan Cake Skill-Share and Free Food," *Indymedia*, May 1, 2010, http://www.indymedia.org.uk/en/2010/05/450210.html; and "Free Vegan Food Giveaway at the AR Spring Gathering," *Indymedia*, March 16, 2010, http://www.indymedia.org.uk/en/2010/03/447611.html.

80 For a report on this larger protest, see "East Midlands Vegan Festival Hits the Streets," *Indymedia*, December 12, 2010, https://www.indymedia.org.uk/en/2010/12/470498.html.

81 Sbicca, "Need to Feed," 826.

82 George Caffentzis and Silvia Federici, "Commons against and beyond Capitalism," *Community Development Journal* 49, no. 1 (2014): 92–105.

83 Feigenbaum, "Resistant Matters," 17.

84 Heynen, "Cooking Up," 1228.

85 For images of workshops, see Patrick Smith, "Vegan Cooking Skillshare July 10," Veggies Catering Campaign, July 29, 2010, http://www.veggies.org.uk/2010/07/vegan-cooking-skillshare-july10/.

86 Megan Warin, "Foucault's Progeny: Jamie Oliver and the Art of Governing Obesity," *Social Theory and Health* 9, no. 1 (2011): 24–40.
87 Barassi, "Ethnographic Cartographies."
88 Dominick, *Animal Liberation*, 13.
89 Carol J. Adams, *The Sexual Politics of Meat: A Feminist Vegetarian Critical Theory* (New York: Continuum, 2000).
90 Mol, "Ontological Politics," 82.
91 Puig de la Bellacasa, "Matters of Care," 97.
92 Stengers, "Experimenting with Refrains," 44.
93 Murphy, *Sick Building Syndrome*, 149.
94 The question of what "trouble" really entails is developed in chapter 5.

4. *Hierarchies of Care*

1 For the mission statement, see the home page of World Day for Laboratory Animals, accessed May 23, 2018, https://worlddayforlaboratoryanimals.org/. For more details on how the day is marked, see "World Day for Laboratory Animals 2018," NAVS: National Anti-Vivisection Society, April 19, 2018, http://www.navs.org.uk/take_action/39/0/4341/.
2 An overview of tensions surrounding activism in the United Kingdom can be found in Carmen McLeod and Pru Hobson-West, "Opening Up Animal Research and Science–Society Relations? A Thematic Analysis of Transparency Discourses in the United Kingdom," *Public Understanding of Science* 25, no. 7 (2016): 791–806.
3 On the early antivivisection movement and responses to it, see Richard French, *Antivivisection and Medical Science in Victorian Society* (Princeton, NJ: Princeton University Press, 1975); and Hilda Kean, "The 'Smooth Cool Men of Science': The Feminist and Socialist Response to Vivisection," *History Workshop Journal* 40, no. 1 (1995): 16–38. On the exclusion of antivivisectionists' concerns from debates about laboratory ethics, see Susan Lederer, "Political Animals: The Shaping of Biomedical Research Literature in Twentieth-Century America," *Isis* 83, no. 1 (1992): 61–79; and Mike Michael and Lynda Birke, "Enrolling the Core Set: The Case of the Animal Experimentation Controversy," *Social Studies of Science* 24, no. 1 (1994): 81–95.
4 The phrase "certain privileged others" is from Lorimer, *Wildlife in the Anthropocene*, 604. In addition to Donna Haraway's arguments in "The Promises of Monsters," these themes are elaborated on in depth within *Modest_Witness@Second_Millennium. FemaleMan©_Meets_OncoMouse™: Feminism and Technoscience* (London: Routledge, 1997).
5 *Fringe* is not intended in a derogatory sense here but is used in reference to groups such as the Safer Medicines Campaign, whose opinions do not accord with the mainstream consensus about animal research. The group, and the broader political role of dissenting knowledge about animal research, is discussed in more depth shortly.
6 "The Felix Campaign," SPEAK: The Voice for the Rights of Animals, SPEAK Campaigns, accessed May 23, 2018, http://www.speakcampaigns.org/felix/felix_campaign.php.

7 For a critique of the power relations of advocacy, see Haraway, "Species Matters."
8 Felix came to the fore in a feature-length BBC documentary, broadcast at the peak of the controversy: Jonathan Wishart, dir., *Monkeys, Rats and Me*, aired November 27, 2006, on BBC 2.
9 Despite common characterizations of appeals to animal subjectivity as irrational or anthropomorphic, this tactic can often prove valuable for activists. In Denmark during the same period, for instance, activists successfully used the tactic of emphasizing primate subjectivity to gain public support. This approach not only succeeded in constructing primate research as morally unacceptable but, in doing so, renegotiated activist identity positions by framing their protest as morally justified. Appeals to emotion, in other words, might not be as irrational as they seem, and instead reveal something important about the political terrain activists have to negotiate—an argument explored in more depth in the next chapter. Lene Koch and Mette N. Svendsen, "Negotiating Moral Value: A Story of Danish Research Monkeys and Their Humans," *Science, Technology, and Human Values* 40, no. 3 (2015): 368–388.
10 *Marginal* is in quotation marks here in recognition that material political disenfranchisement can exist alongside more performative forms of marginalization, wherein normative, conservative perspectives—which are already often hegemonic—are presented as marginal in order to undercut identity claims by genuinely marginalized voices. For an overview of tensions between science and technology studies and post truth, see Des Fitzgerald, "Focus: The Biopolitics of 'Post-truth,'" *Discover Society*, October 4, 2018, https://discoversociety.org/2017/10/04/focus-the-biopolitics-of-post-truth/.
11 Lyle Munro, "Strategies, Action Repertoires and DIY Activism in the Animal Rights Movement," *Social Movement Studies* 4, no. 1 (2005): 75–94.
12 Peter Mason's accessible history of the brown dog affair, for instance, offers an instance of this form of highly visible protest. Although the antivivisection movement originated, and was highly visible, in the Victorian period, one of the most famous early controversies occurred slightly later (between 1903 and 1910). The controversy arose after two women documented alleged abuses to a small brown dog, which they witnessed during their physiology degree at University College London (in which they had purposefully enrolled to gain insight into the treatment of laboratory animals). Their observations were then used in speeches by antivivisectionists (later published as a book, *The Shambles of Science*), which attracted consternation in themselves. The controversy peaked, however, after a statue was erected in the London area of Battersea in 1906. Clashes between medical students, who opposed the statue, and local working-class residents and suffragists seeking to defend it resulted in riots breaking out in 1907. See Peter Mason, *The Brown Dog Affair: The Story of a Monument That Divided a Nation* (London: Two Sevens, 1997); for a reprinted version of the text central to the brown dog affair, see Lizzy Lind-Af-Hageby and Leisa Katherina Schartau, *The Shambles of Science: Extracts from the Diary of Two Students of Physiology* (Miami, FL: HardPress, 2017).
13 For an overview of some of the general debates that emerged about animal activism in the United Kingdom, see Andrew Upton, "'Go On, Get Out There, and Make It

Happen': Reflections on the First Ten Years of Stop Huntingdon Animal Cruelty (SHAC)," *Parliamentary Affairs* 65, no. 1 (2011): 238–254; and John Sorenson, "Constructing Terrorists: Propaganda about Animal Rights," *Critical Studies on Terrorism* 2, no. 2 (2011): 237–256.

14 McLeod and Hobson-West, "Opening Up Animal Research."

15 Pru Hobson-West and Ashley Davies, for instance, frame this shaping of action based on perceptions of what society believes as "societal sentience." See Pru Hobson-West and Ashley Davies, "Societal Sentience: Constructions of the Public in Animal Research Policy and Practice," *Science, Technology, and Human Values* 43, no. 4 (2018): 671–693.

16 The Nexis database was used to locate specific references to this controversy in the UK media; although 150 articles in national newspapers focus directly on SPEAK's activities at Oxford, for the purpose of this chapter I focus on 76 articles from the time when the controversy was at its peak (from March 2004, when construction was announced, to March 2007, after renewed attention to the controversy, following the screening of *Monkeys, Rats and Me*, had died down).

17 Noortje Marres defines an *issue network* as "a variety of political practices that add to and intervene in the representative politics characteristic of national democracies and the international system," adding that the concept has been "taken up by grassroots organisations and individuals in mobilizing around affairs that affect people in their daily lives, from the environment to media ownership and gender issues." What is crucial to this definition is that issue networks are not just composed of actors with complementary perspectives but include the whole range of actors who mobilize around an issue, including those antagonistic toward one another. See Noortje Marres, "Net-Work Is Format Work: Issue Networks and the Sites of Civil Society Politics," in *Reformatting Politics: Information Technology and Global Civil Society*, ed. Jodi Dean, Jon W. Anderson, and Geert Lovink (London: Routledge, 2006), 3–17.

18 The noninnocence of care is a central theme of Puig de la Bellacasa's *Matters of Care*.

19 Gregory Hollin and I provide an overview of arguments about embodied care in Giraud and Hollin, "Care."

20 An agenda for this framework is set out in Puig de la Bellacasa, *Matters of Care*.

21 Latour, "Why Has Critique Run Out of Steam?," 227.

22 It is important to note that this broad characterization of theory has been heavily contested. See, for instance, Benjamin Noys, *The Persistence of the Negative* (Edinburgh: Edinburgh University Press, 2012).

23 Latour, "Why Has Critique Run Out of Steam?," 232.

24 Latour, "Why Has Critique Run Out of Steam?," 237, 246.

25 Puig de la Bellacasa, "Matters of Care," 89.

26 Puig de la Bellacasa, "Matters of Care," 91–92.

27 Puig de la Bellacasa, "Matters of Care," 92.

28 Puig de la Bellacasa, "Matters of Care," 92.

29 Puig de la Bellacasa, "Matters of Care," 88.

30 Puig de la Bellacasa, "Matters of Care," 91.

31 Puig de la Bellacasa, "Matters of Care," 91.

32 For example, Nick Jackson, "Attacks Mount in Animal Battle," *Independent*, February 2, 2006, 6; Michael Seamark and Laura Clark, "Scientists Defy Fanatics to Publicly Back Tests on Animals," *Daily Mail*, February 25, 2006, 16; and Mark Honigsbaum and Alok Jha, "On the Frontline in War over Oxford Animal Laboratory," *Guardian*, January 14, 2006, 10.

33 Aziz was referred to a higher number of times than any other individual, aside from Mel Broughton—a spokesperson for SPEAK—who was mentioned in eighteen articles.

34 Aziz, in Wishart, *Monkeys, Rats and Me*. For an example of the framing of activists, see Rosie Murray West, "Animal Fanatics Attack Drug Firm Director's Home," *Telegraph*, January 30, 2006.

35 For instance, "Resources," SPEAK: The Voice for the Rights of Animals, SPEAK Campaigns, accessed May 26, 2018, http://speakcampaigns.org/resources/; "Wrong Again: The SPEAK Video," SPEAK: The Voice for the Rights of Animals, SPEAK Campaigns, accessed May 26, 2018, http://speakcampaigns.org/the-speak-video/; and "Photo Gallery," SPEAK: The Voice for the Rights of Animals, SPEAK Campaigns, accessed May 26, 2018, http://speakcampaigns.org/photo-gallery.

36 Aziz, in Wishart, *Monkeys, Rats and Me*.

37 Joseph Lister, "Vivisection: Shall We Save a Rabbit and Allow a Man to Die?" *New York Times*, March 7, 1903.

38 The framing offered by the documentary was also taken up by the mainstream media, for example, Stuart Jeffries, "Test Driven," *Guardian*, March 3, 2006, http://www.theguardian.com/uk/2006/mar/04/animalwelfare.highereducation.

39 For a discussion of strategic uses of uncertainty, see Michael Lynch, "The Discursive Production of Uncertainty: The OJ Simpson 'Dream Team' and the Sociology of Knowledge Machine," *Social Studies of Science* 28, nos. 5–6 (1998): 829–869; and Paul N. Edwards, "Global Climate Science, Uncertainty and Politics: Data-Laden Models, Model-Filtered Data," *Science as Culture* 8, no. 4 (1999): 437–472.

40 For an overview of the use of DBS, see Richard G. Bittar et al., "Deep Brain Stimulation for Pain-Relief: A Meta-analysis," *Journal of Clinical Neuroscience* 12, no. 5 (2005): 515–519.

41 John Gardner, "A History of Deep Brain Stimulation: Technological Innovation and the Role of Clinical Assessment Tools," *Social Studies of Science* 43, no. 5 (2013): 707–728; and John Gardner, Gabrielle Samuel, and Clare Williams, "Sociology of Low Expectations: Recalibration as Innovation Work in Biomedicine," *Science, Technology, and Human Values* 40, no. 6 (2015): 998–1021.

42 Gardner, "History," 712.

43 See, for instance, Alim Louis Benabid et al., "Long-Term Suppression of Tremor by Chronic Stimulation of the Ventral Intermediate Thalamic Nucleus," *Lancet* 337, no. 8738 (1991): 403–406; and Gardner, "History," 713.

44 Gardner, "History," 719.

45 See, for instance, Phillip A. Ballard, James W. Tetrud, and J. William Langston, "Permanent Human Parkinsonism due to 1-methyl-4-phenyl-1,2,3,6-tetrahydropyridine (MPTP)," *Neurology* 35, no. 7 (1985): 949–956.

46 R. Stanley Burns et al., "A Primate Model of Parkinsonism," *Proceedings of the National Academy of Sciences of the United States of America* 80, no. 14 (1983): 4546–4550.
47 Gardner, "History," 719.
48 Marina E. Emborg, "Nonhuman Primate Models of Parkinson's Disease," ILAR *Journal* 48, no. 4 (2007): 339–355.
49 Susan H. Fox and Jonathan M. Brotchie, "The MPTP-Lesioned Non-human Primate Models of Parkinson's Disease: Past, Present and Future," *Progress in Brain Research* 184 (2010): 133–157; and Jordi Bové and Céline Perier, "Neurotoxin-Based Models of Parkinson's Disease," *Neuroscience* 211 (2012): 51–76.
50 For the debate over species in early studies, see, for example, Hagai Bergman et al., "The Primate Subthalamic Nucleus: II. Neuronal Activity in the MPTP Model of Parkinsonism," *Journal of Neurophysiology* 72, no. 2 (1991): 507–520. For the methods used to quantify success, see Emborg, "Nonhuman Primate Models."
51 Emborg, "Nonhuman Primate Models," 345.
52 "Parkinson's—the Truth," SPEAK: The Voice for the Rights of Animals, SPEAK Campaigns, accessed May 30, 2018, http://speakcampaigns.org/parkinsons-the-truth/.
53 "Primate Research at Oxford," SPEAK: The Voice for the Rights of Animals, SPEAK Campaigns, accessed May 26, 2018, http://speakcampaigns.org/primate-research-at-oxford/.
54 The reference to Reiss was part of a personal attack on Aziz, in an attempt to challenge claims the researcher had made in *Monkeys, Rats and Me*: "Tipu Aziz Lies Exposed Once Again," SPEAK: The Voice for the Rights of Animals, SPEAK Campaigns, accessed May 26, 2018, http://www.speakcampaigns.org/articles/20070425reiss.php.
55 Kathy Archibald, "BBC Appeal Part I," Safer Medicines Campaign, August 23, 2007, https://safermedicines.org/news_bbcappeal1/. For the BBC's response, "BBC Response," Safer Medicines Campaign, accessed January 25, 2019, https://safermedicines.org/news_bbcprogrespons/.
56 For this specific point, see Susan Leigh Star, "Scientific Work and Uncertainty," *Social Studies of Science* 15, no. 3 (1985): 415. For a sustained exploration of these arguments, see Susan Leigh Star, *Regions of the Mind: Brain Research and the Quest for Scientific Certainty* (Stanford, CA: Stanford University Press, 1989).
57 Star, "Scientific Work and Uncertainty," 393.
58 Many of these perspectives were gathered together on Pro-Test, a site that gained popular media attention after being set up by an Oxford schoolboy in opposition to SPEAK. See "Facts," Pro-Test, accessed May 26, 2018, http://www.pro-test.org.uk/facts.php.
59 Star, "Scientific "Work and Uncertainty," 404–405.
60 Mary Ann Elston, "Women and Anti-vivisection in Victorian England, 1870–1900," in *Vivisection in Historical Perspective*, ed. Nicolaas A. Rupke (New York: Routledge, 1987), 259–287; and Mary Ann Elston, "Attacking the Foundations of Modern Medicine? Antivivisection and the Science of Medicine," in *Challenging Medicine*, 2nd ed., ed. David Kelleher, Jonathan Gabe, and Gareth Williams (London: Routledge, 2006), 196–219.
61 "Ethics," Pro-Test, accessed May 26, 2018, http://www.pro-test.org.uk/facts.php

?lt=a. This approach has been framed as a common tactic for researchers when faced with "fringe science" or skepticism; see Claire Laurier Decoteau and Kelly Underman, "Adjudicating Non-knowledge in Omnibus Autism Proceedings," *Social Studies of Science* 45, no. 4 (2015): 471–500.

62 Wishart, *Monkeys, Rats and Me.*

63 For example, "About Us," Pro-Test, accessed May 26, 2018, http://www.pro-test.org.uk/about.php; "Frequently Asked Questions," Pro-Test, accessed May 26, 2018, http://www.pro-test.org.uk/facts.php?lt=b; and "Facts," Pro-Test, accessed May 26, 2018, http://www.pro-test.org.uk/facts.php.

64 Nicole Nelson, *Model Behaviour: Animal Experiments, Complexity, and the Genetics of Psychiatric Disorders* (Chicago: University of Chicago Press, 2018), 179–180.

65 Haraway, *Modest_Witness.*

66 This process again resonates with Hobson-West and Davies's account of societal sentience. Hobson-West and Davies, "Societal Sentience."

67 An overview of this line of argument is neatly expressed in Matei Candea's work, which elucidates how Isabelle Stengers and Vinciane Despret's work has been used to this end: "Habituating Meerkats and Redescribing Animal Behaviour Science," *Theory, Culture and Society* 30, nos. 7–8 (2013): 105–128.

68 Haraway's work in *When Species Meet* is key here. For a more specific elaboration, see Greenhough and Roe, "Ethics."

69 The notion of felt experience as being integral to research training is elaborated on by Tora Holmberg, "A Feeling for the Animal: On Becoming an Experimentalist," *Society and Animals* 16 (2008): 316–335.

70 Wishart, *Monkeys, Rats and Me.*

71 Wishart, *Monkeys, Rats and Me.*

72 Emborg, "Nonhuman Primate Models."

73 Aziz, in Wishart, *Monkeys, Rats and Me.*

74 See the following chapter for an overview of work that engages with this argument, or, in a different context, Giraud and Hollin, "Care."

75 Aziz, in Wishart, *Monkeys, Rats and Me.*

76 Van Dooren, *Flight Ways*; and Giraud and Hollin, "Care."

77 On animals' inability to consent, see, for example, "Frequently Asked Questions," Pro-Test, accessed May 30, 2018, http://www.pro-test.org.uk/facts.php?lt=b. On animals' ethical incommensurability, see, for example, Jeffries, "Test Driven."

78 Articles noting Singer's appearance include Greg Neale, "Monkey Business," *Independent on Sunday*, December 3, 2006, 63; and Gareth Walsh, "Father of Animal Activism Backs Monkey Testing," *Sunday Times*, November 26, 2006, 12.

79 I have drawn this idea of Disneyfication from the work of Mathias Elrød Madsen and Marie Leth-Espensen, who presented findings from their dissertation at the European Association for Critical Animal Studies conference in 2017. Madsen and Leth-Espensen conducted an extensive discourse analysis of media representations of particularly charismatic animals—Cecil the lion and Marius the giraffe—whose respective deaths (resulting from trophy hunting and euthanasia after zoo overcrowding) had led to public outcry. Within the media, public sentiment was consistently undermined and portrayed as hypocritical due to being grounded in a

Disneyfied conception of these privileged animals. As Madsen and Leth-Espensen suggest, however, the strength of public response should perhaps not be dismissed so swiftly and provides a modest source of hope about the types of ethical reflection that are possible even if the focus remains on charismatic megafauna. This point is developed further in the final chapters of this book. See Mathias Elrød Madsen and Marie Leth-Espensen, "From Public Indignation to Emancipatory Critique" (paper presented at the European Association for Critical Animal Studies Annual Conference, Lund University, Sweden, October 27, 2017).

80 See, for instance, Sergio Sismondo, "Post-truth?," *Social Studies of Science* 47, no. 1 (2017): 3–6.

81 For a helpful overview of debates about uses of uncertainty, see Gregory Hollin, "Autistic Heterogeneity: Linking Uncertainties and Indeterminacies," *Science as Culture* 26, no. 2 (2017): 209–231. For an important instance of publics gaining credibility by engaging with scientific discourse, see Steven Epstein, "The Construction of Lay Expertise: AIDS Activism and the Forging of Credibility in the Reform of Clinical Trials," *Science, Technology, and Human Values* 20, no. 4 (1995): 408–437.

82 In "Care, Laboratory Beagles and Affective Utopia," Hollin and I draw attention to some of the more instrumental ways care has been leveraged historically.

5. *Charismatic Suffering*

1 "Green Hill on Trial," Oppose B&K International, Save the Harlan Beagles, accessed December 20, 2016, http://savetheharlanbeagles.com/bandk/green-hill.html.

2 The brown dog affair (mentioned in the previous chapter, note 12) is perhaps the most famous historical example of how animals can become focusing points; see also Ben Garlick, "Not All Dogs Go to Heaven, Some Go to Battersea: Sharing Suffering and the 'Brown Dog Affair,'" *Social and Cultural Geography* 16, no. 7 (2015): 798–820.

3 These images circulated across social media and were made available through numerous activist sites; a selection of some of the most iconic images from the campaign is provided by Will Potter, "Incredible Photos from Daylight Raid at Green Hill Dog Breeder in Italy," Green Is the New Red, April 30, 2012, http://www.greenisthenewred.com/blog/italy-dog-breeder-rescue-photos/5974/.

4 The translated transcript of the final verdict of Green Hill was archived and made publicly available by a UK anti-beagle-breeding group. "Greenhill Judgement Translation," Oppose B&K Universal, Save the Harlan Beagles, January 23, 2015, http://savetheharlanbeagles.com/bandk/ewExternalFiles/2014-Greenill%20-Judgment-translation-May-15.pdf.

5 Kevin DeLuca, *Image Politics: The New Rhetoric of Environmental Activism* (London: Routledge, 1999).

6 Kari Andén-Papadopoulos, "Citizen Camera-Witnessing: Embodied Political Dissent in the Age of 'Mediated Mass Self-Communication,'" *New Media and Society* 16, no. 5 (2014): 753–769.

7 This concept does not quite map onto Stacy Alaimo's notion of "trans-corporeality," or drawing attention to entanglements between bodies, as it works in a more

analogous way. The body in image events symbolizes shared vulnerability, rather than drawing attention to the ways that particular forces (such as pollutants) have a shared, though differential, impact on bodies. Alaimo, *Exposed*.

8 Directive 2010/63/EU contains more stringent animal welfare regulation and commitments to the three Rs (reduction, refinement, replacement). "Directive 2010/63/EU of the European Parliament and of the Council on the Protection of Animals Used for Scientific Purposes," September 22, 2010, http://eur-lex.europa.eu/legal-content/EN/TXT/?uri=CELEX:32010L0063.

9 Laura Margottini, "Jail Sentences for Staff of Italian Dog Breeding Facility," *Science Magazine*, January 25, 2015, http://www.sciencemag.org/news/2015/01/jail-sentences-staff-italian-dog-breeding-facility.

10 "Greenhill Judgement Translation."

11 For example, Kirk Leech, "The European Animal Research Association Condemns the Guilty Verdicts against the Management of the Green Hill Dog Breeding Facility in Italy," European Animal Research Association, June 16, 2015, http://eara.eu/en/condemantion-statement-on-the-guilty-verdicts-against-the-management-of-the-green-hill-dog-breeding-facility-in-italy/.

12 In making this argument, I am again drawing on Lene Koch and Mette N. Svendsen's analysis of activism in Denmark. Koch and Svendsen, "Negotiating Moral Value."

13 As described in chapter 2, whether digital media can really create space for self-representation has been highly contested. See again Pip Shea, Tanya Notley, and Jean Burgess's overview of frictions and entanglements between activists and media technologies in *Fibreculture*, Shea et al., "Editorial."

14 Zizi Papacharissi, "Affective Publics and Structures of Storytelling: Sentiment, Events and Mediality," *Information, Communication and Society* 19, no. 3 (2016): 308.

15 "Greenhill Judgement Translation," 6.

16 "Greenhill Judgement Translation," 3.

17 For example, Lorimer, *Wildlife in the Anthropocene*.

18 The particular dangers of an affective logic of sympathy, as identified by Jamie Lorimer, are taken up in depth in the next chapter.

19 In making this argument, I am explicitly drawing and elaborating on my existing collaborative work with Gregory Hollin, which I touch on in more depth later in the chapter.

20 Emily Yates-Doerr, "A Reply," Somatosphere, November 13, 2016, http://somatosphere.net/forumpost/a-reply-3.

21 It should be noted that "make kin not babies" has proven especially contentious and been accused of having Malthusian undertones. Haraway denies this and argues that she deliberately worked to avoid this interpretation in *Staying with the Trouble*, with the slogan pointing more toward the necessity of shifting mind-sets and thinking about how to reproduce biodiverse life and forge multispecies communities rather than focusing on the reproduction of the human species. The term nonetheless requires further working through in order to avoid these connotations, which is perhaps not realized within *Staying with the Trouble* itself. See Sophie Lewis, "Cthulu Plays No Role for Me," *Viewpoint*, May 8, 2017, https://www.viewpointmag.com/2017/05/08/cthulhu-plays-no-role-for-me/.

22 Hollin and I engage with these arguments in depth in previous writings about laboratory beagles, for example Giraud and Hollin, "Care"; and Eva Giraud and Gregory Hollin, "Laboratory Beagles and Affective Co-productions of Knowledge," in *Participatory Research in More-than-Human Worlds*, ed. Michelle Bastian, et al. (London: Routledge), 163–177. For an elaboration of the difficulty in creating room for nonhumans to impose their requirements on humans, see Candea, "Habituating Meerkats."

23 Vinciane Despret, "The Body We Care For: Figures of Anthropo-zoo-genesis," *Body and Society* 10, nos. 2–3 (2004): 111–134.

24 Greenhough and Roe, "Ethics."

25 Greenhough and Roe, "Ethics," 57.

26 We develop this application and understanding of Lorimer in further depth in Gregory Hollin and Eva Giraud, "Charisma and the Clinic," *Social Theory and Health* 15, no. 2 (2017): 117–137.

27 Jamie Lorimer, "Charisma," The Multispecies Salon, accessed May 30, 2018, http://www.multispecies-salon.org/charisma/.

28 It should be noted that despite nonhuman charisma being understood as a relational property, ecological charisma in particular is often treated in a more ahistorical sense, related to the fixed anatomical features of particular organisms.

29 Anders Blok, "War of the Whales: Post-sovereign Science and Agonistic Cosmopolitics in Japanese-Global Whaling Assemblages," *Science, Technology, and Human Values* 36, no. 1 (2011): 64.

30 Blok, "War of the Whales," 64.

31 Lorimer, *Wildlife in the Anthropocene*, 58.

32 Van Dooren, *Flight Ways*, 87–124.

33 Lorimer, *Wildlife in the Anthropocene*, 95–96.

34 Hinchliffe et al., "Urban Wild Things."

35 Haraway, *Staying with the Trouble*, 58.

36 Hollin and Giraud, "Charisma and the Clinic," 237.

37 Details of the other experiments in the project are outlined in Roy C. Thompson, *Life-Span Effects of Radiation in the Beagle Dog* (Richland, WA: Department of Energy, Health and Environmental Research, Pacific Northwest Laboratory, 1989).

38 Douglas H. McKelvie and Allen C. Andersen, "Production and Care of Laboratory Beagles," *Journal of the Institute of Animal Technology* 17 (1966): 25–33.

39 Summary of arguments by Allen C. Anderson and Marvin Goldman, in Giraud and Hollin, "Care," 27.

40 Allen Andersen and George Hart, "Kennel Construction and Management in Relation to Longevity Studies in the Dog," *Journal of the American Veterinary Medical Association* 126 (1955): 371.

41 Elizabeth Johnson, "Of Lobsters, Laboratories, and War: Animal Studies and the Temporality of More-than-Human Encounters," *Environment and Planning D: Society and Space* 33, no. 2 (2015): 307.

42 Giraud and Hollin, "Care."

43 Despret, "Body We Care For," 129.

44 Despret, "Body We Care For," 130.

45 Van Dooren, *Flight Ways*, 105.
46 Van Dooren, *Flight Ways*, 103.
47 Van Dooren, *Flight Ways*, 102.
48 Haraway, *When Species Meet*, 22.
49 "Greenhill Judgement Translation," 3–4.
50 "Greenhill Judgement Translation," 35.
51 "Green Hill Beagles," Lega Anti Vivisezione, accessed December 21, 2016, http://www.lav.it/cosa-facciamo/vivisezione/i-beagle-di-green-hill.
52 Sabina Leonelli and Rachel A. Ankeny, "Re-thinking Organisms: The Impact of Databases on Model Organism Biology," *Studies in History and Philosophy of Science Part C: Studies in History and Philosophy of Biological and Biomedical Sciences* 43, no. 1 (2012): 29–36.
53 "Greenhill Judgement Translation," 4.
54 Haraway, *When Species Meet*, 80.
55 "Green Hill on Trial," Oppose B&K International, Save the Harlan Beagles, accessed December 20, 2016, http://savetheharlanbeagles.com/bandk/green-hill.html.
56 Steve Best, "Daring Daytime Dog Liberation at Green Hill Breeders in Italy!," Dr. Steve Best's website, last modified April 28, 2012, https://drstevebest.wordpress.com/2012/04/28/daring-daytime-dog-liberation-at-green-hill-breeders-in-italy.
57 A focus on strengthening researchers' responsiveness to animals' needs, of course, would be anathema from an abolitionist perspective, even though it might be pushed for by welfarists. The point of this chapter, however, is not to delve into this line of argument but to foreground tensions internal to certain forms of situated care ethics, which mean that they are not always as troubling or as nonanthropocentric as they seem. For an overview and argument in favor of abolitionist approaches (and critique of utilitarianism and welfarism), see Corey Lee Wrenn, *A Rational Approach to Animal Rights: Extensions in Abolitionist Theory* (Basingstoke, UK: Palgrave Macmillan, 2015).
58 Mol, "Ontological Politics."
59 As noted above, it is important to recognize that this verdict was not uncontested, with a number of pro-research groups heavily critical of the trial.
60 "Italians Bid to Adopt Beagles Rescued in Cruelty Case," BBC *News*, July 27, 2012, http://www.bbc.co.uk/news/world-europe-19014727.
61 "Directive 2010/63/EU of the European Parliament and of the Council on the Protection of Animals Used for Scientific Purposes," September 22, 2010, http://eur-lex.europa.eu/legal-content/EN/TXT/?uri=CELEX:32010L0063.
62 Peter Singer, *Animal Liberation* (London: Random House, 1995).
63 Hugh Raffles suggests that it is more difficult to engage affectively with certain life-forms (particularly insects) due to their multitudinous nature, which makes the mode of one-to-one encounter-based ethics that has been advocated in more-than-human theoretical contexts difficult to realize. Hugh Raffles, *Insectopedia* (New York: Random House, 2011).

6. *Ambivalent Popularity*

1 *Newsround* is a children's news television program produced by the BBC. Similar stories about the television series, from international and national news outlets, published similar reassurances to "heartbroken" and "devastated" viewers. The David Attenborough–fronted documentary *Planet Earth II* is discussed in more depth later in the chapter. For the full story, see "*Planet Earth II* Turtles Were Saved," *Newsround*, BBC, December 12, 2016, http://www.bbc.co.uk/newsround/38289481.

2 *Influence* here is not meant in the reductive sense of media effects, in terms of straightforwardly shaping behavior, but in relation to consciousness-raising and influence on policy. For instance, three months after episodes of the documentary series *Blue Planet II*, which focused on the problem of seaborne plastic, were screened in November 2017, the issue became highly visible in mainstream media and policy discourse. Imagery used in the program was referred to in 533 UK newspaper articles, in relation to issues surrounding plastic waste (according to media database Nexis), and has since been cited explicitly in local and national policy initiatives seeking to reduce plastic waste. For a summary see Deirdre McKay and Eva Giraud, "Five Ways the Arts Could Help Solve the Plastics Crisis," *The Conversation UK*, January 23, 2018, https://theconversation.com/five-ways-the-arts-could-help-solve-the-plastics-crisis-90136.

3 For example, Sara Ahmed, *The Cultural Politics of Emotion*, 2nd ed. (Edinburgh: Edinburgh University Press, 2014).

4 A helpful elucidation of this is offered in Ruth Holliday and Tracey Potts's discussion of the construction of the "kitsch man," whose taste has been an object of particular derision within aesthetic theory. Ruth Holliday and Tracey Potts, *Kitsch! Cultural Politics and Taste* (Manchester: Manchester University Press, 2012).

5 Gilles Deleuze and Félix Guattari, *A Thousand Plateaus*, trans. Brian Massumi (Minneapolis: University of Minnesota Press, 1987).

6 Haraway, *When Species Meet*, 29–30.

7 Theodor Adorno and Max Horkheimer, *Dialectic of Enlightenment* (London: Verso, 1999), 137.

8 For example, Chris Barker, *Cultural Studies: Theory and Practice* (London: Sage, 2003); and Will Brooker and Deborah Jermyn, *The Audience Studies Reader* (London: Routledge, 2003).

9 Haraway, *When Species Meet*, 67.

10 Lorimer, *Wildlife in the Anthropocene*, 125.

11 I focus here on class, but for sustained arguments about the way aesthetic hierarchies associated with avant-gardism can feed into particular constructions of gender and race, see Marie Thompson, *Beyond Unwanted Sound: Noise, Affect and Aesthetic Moralism* (New York: Bloomsbury, 2017).

12 Mike Goodman et al., "Spectacular Environmentalisms: Media, Knowledge and the Framing of Ecological Politics," *Environmental Communication* 10, no. 6 (2017): 677–688.

13 Guy Debord, *The Society of the Spectacle* (London: Bread and Circuses, 2012).

14 Aside from allegations that the portrayal of political issues within popular culture

automatically depoliticizes them, autonomist Marxist perspectives have tended to see environmental movements themselves as depoliticizing due to distracting from class politics. For an overview and attempt to counter this exclusion, see Sara Nelson and Bruce Braun, "Autonomia in the Anthropocene: New Challenges to Radical Politics," *South Atlantic Quarterly* 116, no. 2 (2017): 223–235.

15 Mike Goodman and Jo Littler, "Celebrity Ecologies: Introduction," *Celebrity Studies* 4, no. 3 (2017): 269–275.

16 Another David Attenborough–fronted documentary series, *Frozen Planet*, for instance, removed an episode focused on climate change in order to sell the series more easily to international television networks. See Andy Bloxham, "BBC Drops *Frozen Planet*'s Climate Change Episode to Sell Show Better Abroad," *Telegraph*, November 11, 2011, http://www.telegraph.co.uk/news/earth/earthnews/8889541/BBC-drops-Frozen-Planets-climate-change-episode-to-sell-show-better-abroad.html.

17 Goodman et al., "Spectacular Environmentalisms," 678.

18 The subtitle of the section, "The Politics of the Popular," gestures toward this field of debate. The title itself is taken from a workshop of the same name that I was involved with at Keele University, led by Holly Kelsall and Wallis Seaton.

19 Rosalind Gill, "Postfeminist Media Culture: Elements of a Sensibility," *European Journal of Cultural Studies* 10, no. 2 (2007): 149.

20 This sort of media text was central to critiques of postfeminist "raunch culture" and a focus on the body as the site of empowerment, as outlined in popular cultural studies books such as Ariel Levy's *Female Chauvinist Pigs: Women and the Rise of Raunch Culture* (New York: Free Press, 2006).

21 In more recent work Gill suggests that postfeminism has become less of a sensibility and more a hegemonic "common sense" way of understanding contemporary gender politics, with problematic consequences for feminism that pushes for structural change. Rosalind Gill, "The Affective, Cultural and Psychic Life of Postfeminism: A Postfeminist Sensibility 10 Years On," *European Journal of Cultural Studies* 20, no. 6 (2017): 606–626.

22 Gill, "Postfeminist Media Culture," 149.

23 Angela McRobbie, "Post-feminism and Popular Culture," *Feminist Media Studies* 4, no. 3 (2004): 255–264.

24 Kaitlynn Mendes, "'Feminism Rules! Now, Where's My Swimsuit?' Re-evaluating Feminist Discourse in Print Media 1968–2008," *Media, Culture and Society* 34, no. 5 (2011): 554–570.

25 Wallis Seaton, "'Doing Her Best with What She's Got': Authorship, Irony, and Mediating Feminist Identities in Lena Dunham's *Girls*," in *Reading Lena Dunham's* Girls, ed. Elizabeth Nash and Imelda Whelehan (Basingstoke, UK: Palgrave MacMillan, 2017), 149–162. For a broader argument, Wallis Seaton, "The Labour of Postfeminist Performance: Postfeminism, Authenticity and Celebrity in Representations of Girlhood on Screen" (PhD diss., Keele University, 2018).

26 As argued by Beverley Skeggs and Helen Wood, eds., *Reacting to Reality Television: Performance, Audience and Value* (Oxon, UK: Routledge, 2012).

27 Haraway, *When Species Meet*, 4.

28 Bastian, "Fatally Confused," 37.
29 Lorimer, *Wildlife in the Anthropocene*, 125.
30 Lorimer, *Wildlife in the Anthropocene*, 127.
31 Lorimer, *Wildlife in the Anthropocene*, 129.
32 Alex Lockwood provides a helpful overview of these discussions in "Graphs of Grief and Other Green Feelings: The Uses of Affect in the Study of Environmental Communication," *Environmental Communication* 10, no. 6 (2016): 734–748.
33 These discussions lie in counterpoint to Haraway's discussion of crittercams, wherein wildlife documentaries include a discussion of the camera techniques and innovations used to see animal worlds and—in doing so—bring the materiality of these technologies to the fore. Haraway, *When Species Meet*, 249–264.
34 Lorimer, *Wildlife in the Anthropocene*, 131.
35 Lorimer, *Wildlife in the Anthropocene*, 135.
36 Ruth Leys, "The Turn to Affect: A Critique," *Critical Inquiry* 37, no. 3 (2011): 434–472; see also Constantina Papoulias and Felicity Callard, "Biology's Gift: Interrogating the Turn to Affect," *Body and Society* 16, no. 1 (2010): 29–56.
37 Edwin S. Porter and/or Jacob Blair Smith, dirs., *Electrocuting an Elephant* (New York: Edison Productions, 1903).
38 For a popular history see Michael Daly, *Topsy: The Startling Story of the Crooked-Tailed Elephant, P.T. Barnum, and the American Wizard, Thomas Edison* (New York: Atlantic Monthly Press, 2013).
39 Nicole Shukin's analysis of *Electrocuting an Elephant* helps to map out key debates. Shukin, *Animal Capital*, 149–161.
40 Rosemary-Claire Collard, "Electric Elephants and the Lively/Lethal Energies of Wildlife Documentary Film," *Area* 48, no. 4 (2016): 472–479. Other arguments that have explored the significance of screening animal death have come from within film studies itself, for example, Akira Mizuta Lippit, "The Death of an Animal," *Film Quarterly* 56, no. 1 (2002): 9–22.
41 For example, Kelly Oliver, "See Topsy 'Ride the Lightning': The Scopic Machinery of Death," *Southern Journal of Philosophy* 50, no. 1 (2012): 74–94; and Elissa Marder, "The Elephant and the Scaffold: Response to Kelly Oliver," *Southern Journal of Philosophy* 50, no. 1 (2012): 95–106.
42 Anat Pick, "Executing Species: Animal Attractions in Thomas Edison and Douglas Gordon," in *The Palgrave Handbook of Posthumanism in Film and Television*, ed. Michael Hauskeller, Curtis D. Carbonell, and Thomas D. Philbeck (Basingstoke, UK: Palgrave Macmillan, 2015), 311–320.
43 Shukin, *Animal Capital*, 141.
44 See Daly, *Topsy*.
45 Maan Barua, "Lively Commodities and Encounter Value," *Environment and Planning D: Society and Space* 34, no. 4 (2016): 725–744.
46 Again, Shukin offers a succinct analysis of the racial dimensions of Topsy's life and execution. Shukin, *Animal Capital*, 154.
47 Isabelle Stengers, "Turtles All the Way Down," in *Power and Invention: Situating Science* (Minneapolis: University of Minnesota Press, 1997), 60–74.

48 Michelle Bastian, "Encountering Leatherbacks in Multispecies Knots of Time," in Bird Rose, van Dooren, and Chrulew, *Extinction Studies*, 149–186; and van Dooren, *Flight Ways*, 12.

49 Shotwell, *Against Purity*, 98.

50 Puig de la Bellacasa, *Matters of Care*, 40.

51 See Simon Cottle, "Producing Nature(s): On the Changing Production Ecology of Natural History TV," *Media, Culture and Society* 26, no. 1 (2004): 81–101.

52 Lorimer, *Wildlife in the Anthropocene*, 132.

53 Quoted in Hannah Furness, "*Planet Earth II* Filmmakers Defy Convention to Save Lost Baby Turtles," *Telegraph*, December 12, 2016, http://www.telegraph.co.uk/news/2016/12/12/bbc-planet-earth-ii-filmmakers-defy-convention-save-lost-baby/.

54 Christopher Hooton, "More Young People Are Watching *Planet Earth 2* than *The X Factor*," *Independent*, December 1, 2016, http://www.independent.co.uk/arts-entertainment/tv/news/planet-earth-2-ii-young-viewers-x-factor-bbc-itv-david-attenborough-vieiwng-figures-ratings-a7449296.html.

55 Martin Hughes-Games, "The BBC's *Planet Earth II* Did Not Help the Natural World," *Guardian*, January 1, 2017, https://www.theguardian.com/commentisfree/2017/jan/01/bbc-planet-earth-not-help-natural-world.

56 Bastian, "Fatally Confused," 41

57 Bastian, "Fatally Confused," 43.

58 Numerous newspapers covered the story. For an overview that documents the BBC's response, see Mary Bowerman, "Baby Turtles Facing Certain Death Saved by 'Planet Earth II' Crew," *USA Today*, December 13, 2016, https://www.usatoday.com/story/news/nation-now/2016/12/13/planet-earth-ii-crew-saved-baby-turtles-certain-death-human-kind-light-turtles-death-cars-drains/95365848/.

59 Quoted in Christopher Hooton, "*Planet Earth 2* Crew Put Every Turtle Hatchling It Saw or Filmed Back in the Sea," *Independent*, December 12, 2016, https://www.independent.co.uk/arts-entertainment/tv/news/planet-earth-2-ii-baby-turtle-hatchlings-scene-conservation-barbados-a7469316.html.

60 Charlotte Bostock, dir., "Hawksbill Turtle Rescue," *Planet Earth II*, BBC, video, 2.53, October 11, 2016, https://www.bbc.co.uk/programmes/p04kccf7.

61 Home page, Official Website of the Barbados Sea Turtle Project, accessed August 23, 2017, http://www.barbadosseaturtles.org/.

62 Matthew Cole and Katie Stewart, *Our Children and Other Animals: The Cultural Construction of Human-Animal Relations in Childhood* (Farnham, UK: Ashgate, 2014).

63 Though Adams's work has proven important, even aside from the aforementioned debates with Haraway, a number of important feminist critiques have emerged of her depiction of pornography; see, for instance, Hamilton, "Sex, Work, Meat."

64 Judith Hampson, for instance, describes how exposés after activists obtained footage of head-injury research led to controversy in the United States during the 1980s (where regulation at the time was not as stringent as in European contexts); see Judith Hampson, "Legislation: A Practical Solution to the Vivisection Dilemma?," in Rupke, *Vivisection in Historical Perspective*, 331–334. In the United Kingdom, the British Union for the Abolition of Vivisection regularly used undercover filming as a tactic, in some instances triggering governmental investigations, for example, Ani-

mal Procedures Committee (APC), *Final Report of the Cambridge/BUAV Working Group*, June 16, 2005, https://webarchive.nationalarchives.gov.uk/20060802125901/http://www.apc.gov.uk/.

65 As described in chapter 4, for instance (see note 12), the brown dog affair was in part triggered by Af Hageby and Schartau's book *The Shambles of Science*, which presented itself as unmasking the animal cruelty that lay behind medical education.

66 Claire Rasmussen, "Pleasure, Pain and Place," in *Critical Animal Geographies: Politics, Intersections and Hierarchies in a Multispecies World*, ed. Kathryn Gillespie and Rosemary-Claire Collard (New York: Routledge, 2015), 54.

67 Rasmussen, "Pleasure, Pain and Place," 54.

68 I use *greenwash* here in the broad sense put forward by scholars who understand it as the process of deliberately misleading consumers about products and processes; see Thomas P. Lyon and A. Wren Montgomery, "The Means and End of Greenwash," *Organization and Environment* 28, no. 2 (2015): 223–249.

69 Kip Andersen and Keegan Kuhn, dirs., *Cowspiracy: The Sustainability Secret* (Los Angeles: AUM Films/First Spark Media, 2014).

70 Lockwood, "Graphs of Grief."

71 Lockwood, "Graphs of Grief," 743.

72 Danny Chivers, "*Cowspiracy*: Stampeding in the Wrong Direction?," *New Internationalist*, February 10, 2016, https://newint.org/blog/2016/02/10/cowspiracy-stampeding-in-the-wrong-direction.

73 Shotwell, *Against Purity*, 125.

74 Shotwell, *Against Purity*, 125.

75 The particular text Shotwell focuses on in making these criticisms is Lierre Keith, *The Vegetarian Myth: Food, Justice, and Sustainability* (Crescent City, CA: PM Press, 2009).

76 This aspect of Shotwell's work, I suggest, is often missed, with the emphasis instead placed on her insistence on the noninnocence of any way of relating.

77 As touched on in the introduction (see note 35), there is a distinct citational politics that obscures complex narratives about vegans' felt experiences. As Carrie Hamilton suggests, this can result in criticisms of figures such as Adams being used to make a straw man out of critical approaches more broadly; see Hamilton, "Sex, Work, Meat."

78 For some helpful analyses of contemporary awareness-raising films; see, for instance, Claire Molloy [Parkinson], "Propaganda, Activism and Environmental Nostalgia," in *Routledge Companion to Cinema and Politics*, ed. Yannis Tzioumakis and Claire Molloy, 139–150 (London: Routledge, 2016).

79 iAnimal (website), Animal Equality, accessed September 8, 2017, http://ianimal.uk/.

80 Mike Goodman, "The Empathy Machine: Virtual Reality, iAnimal and the Techno-Biopolitics of Digital Foodscapes" (paper presented at Digital Food Cultures, Kings College London, July 5, 2017).

81 Lynch plays the role of Luna Lovegood in the *Harry Potter* film franchise. She narrates dir. Animal Equality, "iAnimal: The Dairy Industry in 360°," Animal Equality, Youtube video, 4.46, July 6, 2017, https://www.youtube.com/watch?v=HNIrgmHeI8A.

82 Sean Burch, "New Virtual Reality Series Exposes Cruel Dairy Farm Conditions," *Wrap*, July 6, 2017, http://www.thewrap.com/virtual-reality-animal-cruelty/.

83 Tony Kanal is a member of the band No Doubt. He narrates the iAnimal film: dir. Jose Valle, "iAnimal Pigs," Animal Equality, Youtube video, 10.33, March 1, 2016, https://www.youtube.com/watch?v=A-VMMotnujM.

84 Wolfe, *What Is Posthumanism?*

85 For further reflection on some of the unexpected or inadvertent ways in which the labeling and marketing of animal products has fostered ethical debate, see Mara Miele and John Lever, "Civilizing the Market for Welfare Friendly Products in Europe? The Techno-ethics of the Welfare Quality® Assessment," *Geoforum* 48 (2013): 63–72.

86 Utalkmarketing, "Made by Cows," Anchor Butter, CHI and Partners, Youtube video, 0.39, May 18, 2010, https://www.youtube.com/watch?v=nv1FhC_ascws.

87 Tobias Linné, "Cows on Facebook and Instagram: Interspecies Intimacy in the Social Media Spaces of the Swedish Dairy Industry," *Television and New Media* 17, no. 8 (2016): 719–733.

88 For elaboration on these developments, see Will Davies, *The Happiness Industry: How the Government and Big Business Sold Us Well-Being* (London: Verso Books, 2015).

89 Linné, "Cows on Facebook," 722.

90 In addition to Barua's aforementioned body of work that explores animal labor from a geographical perspective, a number of other thinkers have explored the relationships between Marxism and animals within a range of disciplinary contexts. The issue of *South Atlantic Quarterly* from April 2017 contains a range of articles exploring the relations between animals and autonomist Marxism; Sara Nelson and Bruce Braun, eds., "Autonomia in the Anthropocene," special issue, *South Atlantic Quarterly* 116, no. 2 (2017). For other productive explorations of animals in relation to labor, see Kendra Coulter, *Animals, Work and the Promise of Interspecies Solidarity* (Basingstoke, UK: Palgrave Macmillan, 2015); and Dinesh Wadiwel, *The War against Animals* (Leiden: Brill, 2015).

91 Linné provides an overview of happy meat and new carnist debates in "Cows on Facebook."

92 Tobias Linné and Helena Pedersen, "With Care for Cows and a Love for Milk: Affect and Performance in Dairy Industry Marketing Strategies," in *Meat Culture*, ed. Annie Potts (Leiden: Brill, 2016), 109–128.

Conclusion: An Ethics of Exclusion

1 Again, this argument is intended to refer to multispecies communities rather than focusing solely on the human. The connection between making exclusion visible and creating space for future transformation is why Gregory Hollin has argued for the value of emphasizing Baradian agential cuts in order to support what he describes as an "ethics of transformation." Hollin, "Failing, Hacking, Passing."

2 Barad, *Meeting the Universe Halfway*.

3 Joanna Latimer, "Becoming-Rendered: On Being Caught in between Thresholds," *Threshold* (blog), September 20, 2017, https://thresholdyork.wordpress.com/2017/09/20/becoming-rendered-on-being-caught-in-between-thresholds/.

4 Hollin et al., "(Dis)entangling Barad."

5 Elizabeth Wilson, "Acts against Nature," *Angelaki* 23, no. 1 (2018): 24.
6 Thom van Dooren, "Authentic Crows: Identity, Captivity and Emergent Forms of Life," *Theory, Culture and Society* 33, no. 2 (2016): 43.
7 In particular, van Dooren describes how the decreased vocabulary of captive crows might make it difficult to engage in essential activities such as issuing warnings about predators. Van Dooren, "Authentic Crows," 33.
8 Ginn, "Sticky Lives," 533.
9 Ginn, "Sticky Lives," 541.
10 Mol, "Ontological Politics." For a valuable, critical account of the need to better understand the exclusion central to ontological politics, see Papadopoulos, *Experimental Practice*, 12.
11 The need for clarification on this point was underlined during a helpful discussion after an excellent paper by Florence Chiew, "An Ecology of Ideas with Uexküll and Bateson" (paper presented at Leeds University, May 30, 2018).
12 Haraway, *Staying with the Trouble*, 23. A slightly different articulation of this argument is found in Haraway's *Modest Witness*, 104.
13 Donna Haraway, *The Companion Species Manifesto* (Chicago: Prickly Paradigm Press, 2003), 20.
14 The phrase is from Stengers, "Turtles All the Way Down."
15 Alaimo, *Exposed*, 1.
16 Puig de la Bellacasa, *Matters of Care*, 40. See also this book, chapter 6.
17 Hollin argues that it is necessary to pay close attention to the historical constitution of agential cuts in order to create space for future transformation.
18 The potentially depoliticizing consequences of debunking critical perspectives on the basis that they fail to grasp the hybrid composition of the world has been underlined even by sympathetic commentators, for example Lorimer, *Wildlife in the Anthropocene*, 17.
19 It is important to note that this solution is not uncontentious; even at the time of writing, other interlocutors engaged in debate about whether Freeman's suggestions themselves were an attempt to justify the status quo. See Cathy Levine, "The Tyranny of Tyranny," in Freeman and Levine, *Untying the Knot*, 17–23.
20 I am again here reiterating Rodrigo Nunes's refrain in "Nothing Is What Democracy Looks Like." See chapters 2 and 3.
21 This point has again been developed in Hollin, "Failing, Hacking, Passing."
22 This emphasis on the destructive (though productive) dimensions of exclusion, then, engages with Ginn's call for an ethics oriented around distance and detachment, while also stressing that this line of argument poses difficult questions for relationality itself.
23 As elucidated throughout the book, sometimes a lack of intervention or pluralistic approaches can allow existing relations to continue unabated, along with the realities that the relations bring into being at the expense of alternatives. As argued in chapter 3, therefore, it is dangerous to perceive certain ways of doing things as more troubling than others, as this does not necessarily capture how different approaches relate to preexisting norms and sociotechnical structures.

Bibliography

Adams, Carol J. "An Animal Manifesto: Gender, Identity, and Vegan-Feminism in the Twenty-First Century." Interview by Tom Tyler. *Parallax* 12, no. 1 (2006): 120–128.

Adams, Carol J. *The Sexual Politics of Meat: A Feminist Vegetarian Critical Theory*. New York: Continuum, 2000.

Adorno, Theodor, and Max Horkheimer. *Dialectic of Enlightenment*. London: Verso, 1999.

Ahmed, Sara. *The Cultural Politics of Emotion*, 2nd ed. Edinburgh: Edinburgh University Press, 2014.

Alaimo, Stacy. *Exposed: Environmental Politics and Pleasures in Posthuman Times*. Minneapolis: University of Minnesota Press, 2016.

Allan, Stuart. *Online News: Journalism and the Internet*. Maidenhead, UK: Open University Press, 2006.

Anarchist Teapot. *Feeding the Masses*. N.p.: Active Distribution, n.d.

Andén-Papadopoulos, Kari. "Citizen Camera-Witnessing: Embodied Political Dissent in the Age of 'Mediated Mass Self-Communication.'" *New Media and Society* 16, no. 5 (2014): 753–769.

Andersen, Allen, and George Hart. "Kennel Construction and Management in Relation to Longevity Studies in the Dog." *Journal of the American Veterinary Medical Association* 126 (1955): 366–373.

Andersen, Kip, and Keegan Kuhn, dirs. *Cowspiracy: The Sustainability Secret*. Los Angeles: AUM Films/First Spark Media, 2014.

Armstrong, Franny, and Ken Loach, dirs. *McLibel: Two People Who Wouldn't Say Sorry*. London: Spanner Films, 2005.

Atton, Chris. *Alternative Media*. London: Sage, 2002.

Atton, Chris. "Reshaping Social Movement Media for a New Millennium." *Social Movement Studies* 2, no. 1 (2003): 3–15.

Bach, Jonathan, and David Stark. "Link, Search, Interact: The Co-evolution of NGOs and Interactive Technology." *Theory, Culture and Society* 21, no. 3 (2004): 101–117.

Ballard, Phillip A., James W. Tetrud, and J. William Langston. "Permanent Human Par-

kinsonism due to 1-methyl-4-phenyl-1,2,3,6-tetrahydropyridine (MPTP)." *Neurology* 35, no. 7 (1985): 949–956.
Barad, Karen. "Getting Real: Technoscientific Practices and the Materialization of Reality." *Differences* 10, no. 2 (1998): 87–128.
Barad, Karen. *Meeting the Universe Halfway: Quantum Physics and the Entanglement of Matter and Meaning*. Durham, NC: Duke University Press, 2007.
Barassi, Veronica. *Activism on the Web: Everyday Struggles Against Digital Capitalism*. New York: Routledge, 2015.
Barassi, Veronica. "Ethnographic Cartographies: Social Movements, Alternative Media and the Space of Networks." *Social Movement Studies* 12, no. 1 (2013): 48–62.
Barker, Chris. *Cultural Studies: Theory and Practice*. London: Sage, 2003.
Barua, Maan. "Lively Commodities and Encounter Value." *Environment and Planning D: Society and Space* 34, no. 4 (2016): 725–744.
Bastian, Michelle. "Encountering Leatherbacks in Multispecies Knots of Time." In *Extinction Studies: Stories of Time, Death, and Generations*, edited by Deborah Bird Rose, Thom van Dooren, and Matthew Chrulew, 149–186. New York: Columbia University Press, 2017.
Bastian, Michelle. "Fatally Confused: Telling the Time in the Midst of Ecological Crises." *Environmental Philosophy* 9, no. 1 (2012): 23–48.
Beasts of Burden. London: Active Distribution, 2004.
Beisel, Uli. "Jumping Hurdles with Mosquitos." *Environment and Planning D: Society and Space* 28, no. 1 (2010): 46–49.
Benabid, Alim Louis, P. Pollak, C. Gervason, D. Hoffman, D. M. Gao, M. Hommel, J. E. Perret, and J. de Rougement. "Long-Term Suppression of Tremor by Chronic Stimulation of the Ventral Intermediate Thalamic Nucleus." *Lancet* 337, no. 8738 (1991): 403–406.
Bergman, Hagai, T. Wichmann, B. Karmon, and M. R. DeLong. "The Primate Subthalamic Nucleus: II. Neuronal Activity in the MPTP Model of Parkinsonism." *Journal of Neurophysiology* 72, no. 2 (1991): 507–520.
Best, Steven. "The Rise of Critical Animal Studies: Putting Theory into Action and Animal Liberation into Higher Education." *Journal for Critical Animal Studies* 7, no. 1 (2009): 9–52.
Bird Rose, Deborah, Thom van Dooren, and Matthew Chrulew, eds. *Extinction Studies: Stories of Time, Death, and Generations*. New York: Columbia University Press, 2017.
Bittar, Richard G., Ishani-Kar Purkayastha, Sarah L. Owen, Renee E. Bear, Alex Green, ShouYan Wang, and Tipu Z. Aziz. "Deep Brain Stimulation for Pain-Relief: A Meta-analysis." *Journal of Clinical Neuroscience* 12, no. 5 (2005): 515–519.
Blok, Anders. "War of the Whales: Post-sovereign Science and Agonistic Cosmopolitics in Japanese-Global Whaling Assemblages." *Science, Technology, and Human Values* 36, no. 1 (2011): 55–81.
Bové, Jordi, and Céline Perier. "Neurotoxin-Based Models of Parkinson's Disease." *Neuroscience* 211 (2012): 51–76.
Braidotti, Rosi. *The Posthuman*. London: Polity, 2012.
Brooker, Will, and Deborah Jermyn. *The Audience Studies Reader*. London: Routledge, 2003.

Buller, Henry. "Animal Geographies I." *Progress in Human Geography* 38, no. 2 (2014): 308–318.

Buller, Henry. "Animal Geographies III: Ethics." *Progress in Human Geography* 40, no. 3 (2016): 422–430.

Burns, R. Stanley, Chuang C. Chiueh, Sanford P. Markey, Michael H. Ebert, David M. Jacobwitz, and Irwin J. Kopin. "A Primate Model of Parkinsonism." *Proceedings of the National Academy of Sciences of the United States of America* 80, no. 14 (1983): 4546–4550.

Caffentzis, George, and Silvia Federici. "Commons against and beyond Capitalism." *Community Development Journal* 49, no. 1 (2014): 92–105.

Callard, Felicity, and Des Fitzgerald. *Rethinking Interdisciplinarity across the Social Sciences and Neurosciences*. Basingstoke, UK: Palgrave Macmillan, 2016.

Candea, Matei. "Habituating Meerkats and Redescribing Animal Behaviour Science." *Theory, Culture and Society* 30, nos. 7–8 (2013): 105–128.

Carpentier, Nico, Peter Dahlgren, and Francesca Pasquali. "Waves of Media Democratization: A Brief History of Contemporary Participatory Practices in the Media Sphere." *Convergence: The International Journal of Research into New Media Technologies* 19, no. 3 (2013): 287–294.

Castells, Manuel. *The Power of Identity*. Malden, MA: Blackwell, 1997.

Castree, Noel, Catherine Nash, Neil Badmington, Bruce Braun, Jonathon Murdoch, and Sarah Whatmore. "Mapping Posthumanism: An Exchange." *Environment and Planning A: Economy and Space* 36, no. 8 (2004): 1341–1363.

Chatterton, Paul, and Jenny Pickerill. "Everyday Activism and the Transitions towards Post-capitalist Worlds." *Transactions of the Institute of British Geographers* 35, no. 4 (2010): 475–490.

Chiew, Florence. "An Ecology of Ideas with Uexküll and Bateson." Paper presented at Leeds University, May 30, 2018.

Clark, Jonathan L. "Uncharismatic Invasives." *Environmental Humanities* 6, no. 1 (2015): 29–52.

Cleaver, Harry. "The Chiapas Uprising and the Future of Class Struggle in the New World Order." February 14, 1994. http://www.eco.utexas.edu/facstaff/Cleaver/chiapasuprising.html.

Cole, Matthew, and Katie Stewart. *Our Children and Other Animals: The Cultural Construction of Human-Animal Relations in Childhood*. Farnham, UK: Ashgate, 2014.

Colebrook, Claire. *Death of the PostHuman: Essays on Extinction*. Vol. 1. Ann Arbor, MI: Open Humanities Press, 2014.

Collard, Rosemary-Claire. "Electric Elephants and the Lively/Lethal Energies of Wildlife Documentary Film." *Area* 48, no. 4 (2016): 472–479.

Collard, Rosemary-Claire. "Putting Animals Back Together, Taking Commodities Apart." *Annals of the Association of American Geographers* 104, no. 1 (2015): 151–165.

Cottle, Simon. "Producing Nature(s): On the Changing Production Ecology of Natural History TV." *Media, Culture and Society* 26, no. 1 (2004): 81–101.

Couldry, Nick, and Andreas Hepp. *The Mediated Construction of Reality*. London: Polity, 2016.

Coulter, Kendra. *Animals, Work and the Promise of Interspecies Solidarity*. Basingstoke, UK: Palgrave Macmillan, 2015.

CounterSpin Collective. "Media, Movement(s) and Public Image(s): Counterspinning in Scotland." In *Shut Them Down! The G8, Gleneagles 2005 and the Movement of Movements*, edited by David Harvie, Kier Milburn, Ben Trott, and David Watts, 321–333. Leeds, UK: Dissent!; Brooklyn, NY: Autonomedia, 2005.

Cudworth, Erika. *Social Lives with Other Animals: Tales of Sex, Death and Love*. Basingstoke, UK: Palgrave Macmillan, 2011.

Daly, Michael. *Topsy: The Startling Story of the Crooked-Tailed Elephant, P.T. Barnum, and the American Wizard, Thomas Edison*. New York: Atlantic Monthly Press, 2013.

Davies, Will. *The Happiness Industry: How the Government and Big Business Sold Us Well-Being*. London: Verso Books, 2015.

Dean, Jodi. *Democracy and Other Neoliberal Fantasies: Communicative Capitalism and Left Politics*. Durham, NC: Duke University Press, 2009.

Debord, Guy. *The Society of the Spectacle*. London: Bread and Circuses, 2012.

de Certeau, Michel. *The Practice of Everyday Life*. Translated by Steven Rendall. Berkeley: University of California Press, 1984.

Decoteau, Claire Laurier, and Kelly Underman. "Adjudicating Non-knowledge in Omnibus Autism Proceedings." *Social Studies of Science* 45, no. 4 (2015): 471–500.

Deleuze, Gilles, and Félix Guattari. *A Thousand Plateaus*. Translated by Brian Massumi. Minneapolis: University of Minnesota Press, 1987.

Della Porta, Donatella, ed. *Global Justice Movement: Cross-national and Transnational Perspectives*. London: Routledge, 2015.

Della Porta, Donatella. "Making the Polis: Social Forums and Democracy in the Global Justice Movement." *Mobilization: An International Quarterly* 10, no. 1 (2005): 73–94.

DeLuca, Kevin. *Image Politics: The New Rhetoric of Environmental Activism*. London: Routledge, 1999.

Demeritt, Daniel. "Science Studies, Climate Change and the Prospects for Constructivist Critique." *Economy and Society* 35, no. 3 (2006): 453–479.

Despret, Vinciane. "The Body We Care For: Figures of Anthropo-zoo-genesis." *Body and Society* 10, nos. 2–3 (2004): 111–134.

Despret, Vinciane. "Responding Bodies and Partial Affinities in Human-Animal Worlds." *Theory, Culture and Society* 30, nos. 7–8 (2013): 51–76.

Despret, Vinciane. *What Would Animals Say If We Asked the Right Questions?* Translated by Brett Buchanan. Minneapolis: University of Minnesota Press, 2016.

Dominick, Brian. *Animal Liberation and Social Revolution: A Vegan Perspective on Anarchism or an Anarchist Perspective on Veganism*, 2nd ed. London: Active Distribution, 1998.

Downey, John, and Natalie Fenton. "New Media, Counter-publicity and the Public Sphere." *New Media and Society* 5, no. 2 (2003): 185–202.

Edwards, Paul N. "Infrastructure and Modernity: Force, Time, and Social Organization in the History of Sociotechnical Systems." In *Modernity and Technology*, edited by Thomas J. Misa, Philip Brey, and Andrew Feenberg, 185–226. Cambridge, MA: MIT Press, 2003.

Edwards, Paul N. "Global Climate Science, Uncertainty and Politics: Data Laden Models, Model-Filtered Data." *Science as Culture* 8, no. 4 (1999): 437–472.

Eisenman, Stephen F. "The Real 'Swinish Multitude.'" *Critical Inquiry* 42, no. 2 (2016): 339–373.

Elston, Mary Ann. "Attacking the Foundations of Modern Medicine? Antivivisection and the Science of Medicine." In *Challenging Medicine*, 2nd ed., edited by David Kelleher, Jonathan Gabe, and Gareth Williams, 196–219. London: Routledge, 2006.
Elston, Mary Ann. "Women and Anti-vivisection in Victorian England, 1870–1900." In *Vivisection in Historical Perspective*, edited by Nicolaas A. Rupke, 259–287. New York: Routledge, 1987.
Emborg, Marina E. "Nonhuman Primate Models of Parkinson's Disease." *ILAR Journal* 48, no. 4 (2007): 339–355.
Epstein, Steven. "The Construction of Lay Expertise: AIDS Activism and the Forging of Credibility in the Reform of Clinical Trials." *Science, Technology, and Human Values* 20, no. 4 (1995): 408–437.
Fanon, Frantz. *The Wretched of the Earth*. Translated by Constance Farrington. St Ives, UK: Penguin, 1967.
Feigenbaum, Anna. "Resistant Matters: Tear Gas, Tents and the 'Other Media' of Occupy." *Communication and Critical/Cultural Studies* 11, no. 1 (2014): 15–24.
Feigenbaum, Anna, Fabien Frenzel, and Patrick McCurdy. *Protest Camps*. London: Zed, 2013.
Fish, Adam. "Mirroring the Videos of Anonymous: Cloud Activism, Living Networks, and Political Mimesis." *Fibreculture* 26 (2015): 85–107.
Fitzgerald, Des. "The Affective Labour of Autism Neuroscience." *Subjectivity* 6 (2013): 131–152.
Fitzgerald, Des, and Felicity Callard. "Social Science and Neuroscience beyond Interdisciplinarity: Experimental Entanglements." *Theory, Culture and Society* 32, no. 1 (2015): 3–32.
Fox, Susan H., and Jonathan M. Brotchie. "The MPTP-Lesioned Non-human Primate Models of Parkinson's Disease: Past, Present and Future." *Progress in Brain Research* 184 (2010): 133–157.
Franklin, Sarah. "Staying with the Manifesto: An Interview with Donna Haraway." *Theory, Culture and Society* 34, no. 4 (2017): 49–63.
Freeman, Jo. "The Tyranny of Structurelessness." In *Untying the Knot: Feminism, Anarchism and Organisation*, edited by Jo Freeman and Cathy Levine, 5–16. Whitechapel, London: Dark Star and Rebel Press, 1984.
French, Richard. *Antivivisection and Medical Science in Victorian Society*. Princeton, NJ: Princeton University Press, 1975.
Frenzel, Fabian, Steffan Böhm, Pennie Quinton, André Spicer, Sian Sullivan, and Zoe Young. "Comparing Alternative Media in North and South." *Environment and Planning A: Economy and Space* 43, no. 5 (2012): 1173–1189.
Frizzell, Deborah, and Harry J. Weil, curators. *Women in the Wilderness*. Exhibition at Wave Hill, Glyndor Gallery, New York, April 9–July 9, 2017.
Fuller, Matthew, and Andrew Goffey. *Evil Media*. Cambridge, MA: MIT Press, 2012.
Garcelon, Marc. "The 'Indymedia' Experiment: The Internet as Movement Facilitator against Institutional Control." *Convergence: The International Journal of Research into New Media Technologies* 12, no. 1 (2006): 55–82.
Gardner, John. "A History of Deep Brain Stimulation: Technological Innovation and the Role of Clinical Assessment Tools." *Social Studies of Science* 43, no. 5 (2013): 707–728.

Gardner, John, Gabrielle Samuel, and Clare Williams. "Sociology of Low Expectations: Recalibration as Innovation Work in Biomedicine." *Science, Technology, and Human Values* 40, no. 6 (2015): 998–1021.
Garlick, Ben. "Not All Dogs Go to Heaven, Some Go to Battersea: Sharing Suffering and the 'Brown Dog Affair.'" *Social and Cultural Geography* 16, no. 7 (2015): 798–820.
Gill, Rosalind. "The Affective, Cultural and Psychic Life of Postfeminism: A Postfeminist Sensibility 10 Years On." *European Journal of Cultural Studies* 20, no. 6 (2017): 606–626.
Gill, Rosalind. "Postfeminist Media Culture: Elements of a Sensibility." *European Journal of Cultural Studies* 10, no. 2 (2007): 147–166.
Ginn, Franklin. "Sticky Lives: Slugs, Detachment and More-than-Human Ethics in the Garden." *Transactions of the Institute of British Geographers* 39, no. 4 (2014): 532–544.
Giraud, Eva. "Displacement, 'Failure' and Friction: Tactical Interventions in the Communication Ecologies of Anti-capitalist Food Activism." In *Digital Food Activism*, edited by Tanja Schneider, Karin Eli, Catherine Dolan, and Stanley Ulijaszek, 130–150. New York: Routledge, 2018.
Giraud, Eva. "Feminist Praxis, Critical Theory and Informal Hierarchies." *Journal of Feminist Scholarship* 7, no. 1 (2015): 43–60.
Giraud, Eva. "Has Radical Participatory Online Media Really 'Failed'? Indymedia and Its Legacies." *Convergence: The International Journal of Research into New Media Technologies* 20, no. 4 (2014): 419–437.
Giraud, Eva. "Practice as Theory: Learning from Food Activism and Performative Protest." In *Critical Animal Geographies: Politics, Intersections and Hierarchies in a Multispecies World*, edited by Kathryn Gillespie and Rosemary-Claire Collard, 36–53. New York: Routledge, 2015.
Giraud, Eva. "Veganism as Affirmative Biopolitics." *PhaenEx* 8, no. 2 (2014): 47–79.
Giraud, Eva, and Gregory Hollin. "Care, Laboratory Beagles and Affective Utopia." *Theory, Culture and Society* 33, no. 4 (2016): 27–49.
Giraud, Eva, and Gregory Hollin. "Laboratory Beagles and Affective Co-productions of Knowledge." In *Participatory Research in More-than-Human Worlds*, edited by Michelle Bastian, Owain Jones, Niamh Moore, and Emma Roe, 163–177. London: Routledge, 2017.
Giraud, Eva, Gregory Hollin, Tracey Potts, and Isla Forsyth. "A Feminist Menagerie." *Feminist Review* 118, no. 1 (2018): 61–79.
Goodman, Mike. "The Empathy Machine: Virtual Reality, iAnimal and the Techno-Biopolitics of Digital Foodscapes." Paper presented at Digital Food Cultures, Kings College London, July 5, 2017.
Goodman, Mike, and Jo Littler. "Celebrity Ecologies: Introduction." *Celebrity Studies* 4, no. 3 (2017): 269–275.
Goodman, Mike, Jo Littler, Dan Brockington, and Maxwell Boycoff. "Spectacular Environmentalisms: Media, Knowledge and the Framing of Ecological Politics." *Environmental Communication* 10, no. 6 (2017): 677–688.
Graeber, David. *Direct Action: An Ethnography*. Edinburgh: AK Press, 2009.
Greenhough, Beth, and Emma Roe. "Ethics, Space, and Somatic Sensibilities: Comparing Relationships between Scientific Researchers and Their Human and Ani-

mal Experimental Subjects." *Environment and Planning D: Society and Space* 29, no. 1 (2011): 47–66.
Gruen, Lori. *Entangled Empathy: An Alternative Ethic for Our Relationships with Animals.* Brooklyn, NY: Lantern Books, 2015.
Guattari, Félix. *The Three Ecologies.* Translated by Ian Pindar and Paul Sutton. London: Continuum, 2008.
Hamilton, Carrie. "Sex, Work, Meat: The Feminist Politics of Veganism." *Feminist Review* 114, no. 1 (2016): 112–129.
Hampson, Judith. "Legislation: A Practical Solution to the Vivisection Dilemma?" In *Vivisection in Historical Perspective*, edited by Nicolaas A. Rupke, 331–334. New York: Routledge, 1987.
Hands, Joss. *@ Is for Activism: Dissent, Resistance and Rebellion in a Digital Culture.* London: Pluto, 2010.
Hands, Joss. "Civil Society, Cosmopolitics and the Net: The Legacy of 15 February 2003." *Information, Communication and Society* 9, no. 2 (2006): 225–243.
Hands, Joss, Greg Elmer, and Ganaele Langlois, eds. "Platform Politics." Special issue, *Culture Machine* 14 (2013). https://monoskop.org/images/c/ce/Culture_Machine_Vol_14_Platform_Politics.pdf.
Haran, Joan. "Instantiating Imaginactivism: Le Guin's The Dispossessed as Inspiration." ADA: *A Journal of Gender, New Media & Technology* 12 (2017). Accessed February 2, 2019. https://adanewmedia.org/2017/10/issue12-haran/.
Haraway, Donna. *The Companion Species Manifesto.* Chicago, IL: Prickly Paradigm Press, 2003.
Haraway, Donna. "A Cyborg Manifesto: Science, Technology, and Socialist-Feminism in the Late Twentieth Century." In *Simians, Cyborgs, and Women*, 127–148. London: Routledge, 1991.
Haraway, Donna. *Modest_Witness@Second_Millennium. FemaleMan©_Meets_OncoMouse™: Feminism and Technoscience.* London: Routledge, 1997.
Haraway, Donna. "The Promises of Monsters: A Regenerative Politics for Inappropriate/d Others." In *Cultural Studies*, edited by Lawrence Grossberg, Cary Nelson, and Paula Treichler, 295–337. New York: Routledge, 1992.
Haraway, Donna. "Species Matters, Humane Advocacy: In the Promising Grip of Earthly Oxymorons." In *Species Matters: Humane Advocacy and Cultural Theory*, edited by Marianne DeKoven and Michael Lundblad, 17–26. New York: Columbia University Press, 2011.
Haraway, Donna. *Staying with the Trouble: Making Kin in the Chthulucene.* Durham, NC: Duke University Press, 2016.
Haraway, Donna. *When Species Meet.* Minneapolis: University of Minnesota Press, 2008.
Harding, Eloise. "Conceptualising Horizontal Politics." PhD diss., University of Nottingham, 2012.
Harding, Sandra. *The Postcolonial Science and Technology Studies Reader.* Durham, NC: Duke University Press, 2011.
Harper, Breeze A. *Sistah Vegan: Food, Identity, Health, and Society; Black Female Vegans Speak.* Brooklyn, NY: Lantern Books, 2010.

Harvie, David, Keir Milburn, Ben Trott, and David Watts, eds. *Shut Them Down! The G8, Gleneagles 2005 and the Movement of Movements*. Leeds, UK: Dissent!; Brooklyn, NY: Autonomedia, 2005.

Hecht, Susanna, and Alexander Cockburn. *The Fate of the Forest: Developers, Destroyers and Defenders of the Amazon*. Chicago: University of Chicago Press, 1990.

Heynen, Nik. "Cooking Up Non-violent Civil-Disobedient Direct Action for the Hungry." *Urban Studies* 47, no. 6 (2010): 1225–1240.

Hinchliffe, Steve, Matthew B. Kearnes, Monica Degan and Sarah Whatmore. "Urban Wild Things: A Cosmopolitical Experiment." *Environment and Planning D: Society and Space* 23, no. 5 (2005): 643–658.

Hobson-West, Pru, and Ashley Davies. "Societal Sentience: Constructions of the Public in Animal Research Policy and Practice." *Science, Technology, and Human Values* (2018): 671–693.

Holliday, Ruth, and Tracey Potts. *Kitsch! Cultural Politics and Taste*. Manchester: Manchester University Press, 2012.

Hollin, Gregory. "Autistic Heterogeneity: Linking Uncertainties and Indeterminacies." *Science as Culture* 26, no. 2 (2017): 209–231.

Hollin, Gregory. "Failing, Hacking, Passing: Autism, Entanglement, and the Ethics of Transformation." *BioSocieties* 12, no. 4 (2017): 611–633.

Hollin, Gregory, Isla Forsyth, Eva Giraud, and Tracey Potts. "(Dis)entangling Barad: Materialisms and Ethics." *Social Studies of Science* 47, no. 6 (2017): 918–941.

Hollin, Gregory, and Eva Giraud. "Charisma and the Clinic." *Social Theory and Health* 15, no. 2 (2017): 117–137.

Holloway, John, and Elena Peláez. *Zapatista! Reinventing Revolution in Mexico*. London: Pluto, 1997.

Holmberg, Tora. "A Feeling for the Animal: On Becoming an Experimentalist." *Society and Animals* 16 (2008): 316–335.

Hulme, Alison. *On the Commodity Trail: The Journey of a Bargain Store Product from East to West*. London: Bloomsbury, 2015.

Ippolita, Geert Lovink, and Ned Rossiter. "The Digital Given: 10 Web 2.0 Theses." *Fibreculture* 14 (2009). Accessed February 1, 2019. http://fourteen.fibreculturejournal.org/fcj-096-the-digital-given-10-web-2-0-theses/.

Irni, Sari. "The Politics of Materiality: Affective Encounters in a Transdisciplinary Debate." *European Journal of Women's Studies* 20, no. 4 (2013): 347–360.

Johnson, Elizabeth. "Of Lobsters, Laboratories and War: Animal Studies and the Temporality of More-than-Human Encounters." *Environment and Planning D: Society and Space* 33, no. 2 (2015): 296–313.

Juris, Jeffrey. *Networking Futures: The Movements against Corporate Globalization*. Durham, NC: Duke University Press, 2007.

Juris, Jeffrey. "Performing Politics: Image, Embodiment and Affective Solidarity during Anti-corporate Globalization Protests." *Ethnography* 9, no. 1 (2008): 61–97.

Kahn, Richard, and Doug Kellner. "New Media and Internet Activism: From the 'Battle of Seattle' to Blogging." *New Media and Society* 6, no. 1 (2004): 87–95.

Kean, Hilda. "The 'Smooth Cool Men of Science': The Feminist and Socialist Response to Vivisection." *History Workshop Journal* 40, no. 1 (1995): 16–38.

Keith, Lierre. *The Vegetarian Myth: Food, Justice, and Sustainability*. Crescent City, CA: PM Press, 2009.

Klein, Naomi. *No Logo*. London: Flamingo, 2000.

Koch, Lene, and Mette N. Svendsen. "Negotiating Moral Value: A Story of Danish Research Monkeys and Their Humans." *Science, Technology, and Human Values* 40, no. 3 (2015): 368–388.

Latimer, Joanna. "Being Alongside: Rethinking Relations amongst Different Kinds." *Theory, Culture and Society* 30, nos. 7–8 (2013): 77–104.

Latimer, Joanna, and Mara Miele. "Naturecultures: Science, Affect and the Nonhuman." *Theory, Culture and Society* 30, nos. 7–8 (2013): 5–31.

Latour, Bruno. *We Have Never Been Modern*. Cambridge, MA: Harvard University Press, 1993.

Latour, Bruno. "Why Has Critique Run Out of Steam? From Matters of Fact to Matters of Concern." *Critical Inquiry* 30, no. 2 (2004): 225–248.

Law, John, and Annemarie Mol. *Complexities: Social Studies of Knowledge Practices*. Durham, NC: Duke University Press, 2002.

Lederer, Susan. "Political Animals: The Shaping of Biomedical Research Literature in Twentieth-Century America." *Isis* 83, no. 1 (1992): 61–79.

LeDuff, Charlie. "At the Slaughterhouse Some Things Never Die." In *Zoontologies: The Question of the Animal*, edited by Cary Wolfe, 183–198. Minneapolis: University of Minnesota Press, 2003.

Leonelli, Sabina, and Rachel A. Ankeny. "Re-thinking Organisms: The Impact of Databases on Model Organism Biology." *Studies in History and Philosophy of Science Part C: Studies in History and Philosophy of Biological and Biomedical Sciences* 43, no. 1 (2012): 29–36.

Levine, Cathy. "The Tyranny of Tyranny." In *Untying the Knot: Feminism, Anarchism and Organisation*, by Jo Freeman and Cathy Levine, 17–23. Whitechapel, London: Dark Star and Rebel Press, 1984.

Levy, Ariel. *Female Chauvinist Pigs: Women and the Rise of Raunch Culture*. New York: Free Press, 2006.

Lewis, Sophie. "Cthulhu Plays No Role for Me." *Viewpoint*, May 8, 2017. https://www.viewpointmag.com/2017/05/08/cthulhu-plays-no-role-for-me/.

Leys, Ruth. "The Turn to Affect: A Critique." *Critical Inquiry* 37, no. 3 (2011): 434–472.

Lievrouw, Leah. *Alternative and Activist New Media*. London: Polity, 2011.

Lind-Af-Hageby, Lizzy, and Leisa Katherina Schartau. *The Shambles of Science: Extracts from the Diary of Two Students of Physiology*. Miami, FL: HardPress, 2017.

Linné, Tobias. "Cows on Facebook and Instagram: Interspecies Intimacy in the Social Media Spaces of the Swedish Dairy Industry." *Television and New Media* 17, no. 8 (2016): 719–733.

Linné, Tobias, and Helena Pedersen. "With Care for Cows and a Love for Milk: Affect and Performance in Dairy Industry Marketing Strategies." In *Meat Culture*, edited by Annie Potts, 109–128. Leiden: Brill, 2016.

Lippit, Akira Mizuta. "The Death of an Animal." *Film Quarterly* 56, no. 1 (2002): 9–22.

Lockwood, Alex. "Graphs of Grief and Other Green Feelings: The Uses of Affect in the Study of Environmental Communication." *Environmental Communication* 10, no. 6 (2016): 734–748.

Lorimer, Jamie. "Multinatural Geographies for the Anthropocene." *Progress in Human Geography* 36, no. 5 (2012): 593–612.
Lorimer, Jamie. *Wildlife in the Anthropocene: Conservation after Nature*. Minneapolis: University of Minnesota Press, 2015.
Lynch, Michael. "The Discursive Production of Uncertainty: The OJ Simpson 'Dream Team' and the Sociology of Knowledge Machine." *Social Studies of Science* 28, nos. 5–6 (1998): 829–869.
Lynn, William. "Animals, Ethics, and Geography." In *Animal Geographies: Place, Politics, and Identity in the Nature-Culture Borderlands*, edited by Jennifer R. Wolch and Jody Emel, 280–298. New York: Verso, 1998.
Lyon, Thomas P., and A. Wren Montgomery. "The Means and End of Greenwash." *Organization and Environment* 28, no. 2 (2015): 223–249.
Madsen, Mathias Elrød, and Marie Leth-Espensen. "From Public Indignation to Emancipatory Critique." Paper presented at the European Association for Critical Animal Studies Annual Conference, Lund University, Sweden, October 27, 2017.
Marder, Elissa. "The Elephant and the Scaffold: Response to Kelly Oliver." *Southern Journal of Philosophy* 50, no. 1 (2012): 95–106.
Marres, Noortje. "Net-Work Is Format Work: Issue Networks and the Sites of Civil Society Politics." In *Reformatting Politics: Information Technology and Global Civil Society*, edited by Jodi Dean, Jon W. Anderson, and Geert Lovink, 3–17. London: Routledge, 2006.
Martin, Aryn, Natasha Myers, and Ana Viseu. "The Politics of Care in Technoscience." *Social Studies of Science* 45, no. 5 (2015): 625–641.
Mason, Peter. *The Brown Dog Affair: The Story of a Monument that Divided the Nation*. London: Two Sevens, 1997.
McHenry, Keith. *Hungry for Peace: How You Can Help End Poverty and War with Food Not Bombs*. Tucson, AZ: See Sharp, 2012.
McKay, Deirdre. "Subversive Plasticity." With Padmapani Perez, Ruel Bimuyag, and Raja Shanti Bonnevie. In *The Social Life of Materials: Studies in Materials and Society*, edited by Suzanne Küchler and Adam Drazin, 175–192. London: Bloomsbury Academic, 2015.
McKelvie, Douglas H., and Allen C. Andersen. "Production and Care of Laboratory Beagles." *Journal of the Institute of Animal Technology* 17 (1966): 25–33.
McLeod, Carmen, and Pru Hobson-West. "Opening Up Animal Research and Science–Society Relations? A Thematic Analysis of Transparency Discourses in the United Kingdom." *Public Understanding of Science* 25, no. 7 (2016): 791–806.
McRobbie, Angela. "Post-feminism and Popular Culture." *Feminist Media Studies* 4, no. 3 (2004): 255–264.
Mendes, Kaitlynn. "'Feminism Rules! Now, Where's My Swimsuit?' Re-evaluating Feminist Discourse in Print Media, 1968–2008." *Media, Culture and Society* 34, no. 5 (2011): 554–570.
Mercea, Dan, Laura Iannelli, and Brian D. Loader. "Protest Communication Ecologies." *Information, Communication and Society* 19, no. 3 (2016): 279–289.
Mercea, Dan, Laura Iannelli, and Brian D. Loader, eds. "Protest Communication Ecologies." Special issue, *Information, Communication and Society* 19, no. 3 (2016).

Michael, Mike, and Lynda Birke. "Enrolling the Core Set: The Case of the Animal Experimentation Controversy." *Social Studies of Science* 24, no. 1 (1994): 81–95.

Miele, Mara, and John Lever. "Civilizing the Market for Welfare Friendly Products in Europe? The Techno-ethics of the Welfare Quality® Assessment." *Geoforum* 48 (2013): 63–72.

Mitchell, Don, and Nik Heynen. "The Geography of Survival and the Right to the City." *Urban Geography* 30, no. 6 (2009): 611–632.

Mol, Annemarie. *The Body Multiple: Ontology in Medical Practice*. Durham, NC: Duke University Press, 2002.

Mol, Annemarie. "Ontological Politics: A Word and Some Questions." In *Actor Network Theory and After*, edited by John Law and John Hassard, 74–89. Oxford: Blackwell, 1999.

Molloy, Claire. "Propaganda, Activism and Environmental Nostalgia." In *Routledge Companion to Cinema and Politics*, edited by Yannis Tzioumakis and Claire Molloy, 139–150. London: Routledge, 2016.

Munro, Lyle. "Strategies, Action Repertoires and DIY Activism in the Animal Rights Movement." *Social Movement Studies* 4, no. 1 (2005): 75–94.

Murphy, Michelle. *Sick Building Syndrome and the Problem of Uncertainty*. Durham, NC: Duke University Press, 2006.

Nagy, Kelsi, and Phillip David Johnson. *Trash Animals: How We Live with Nature's Filthy, Feral, Invasive, and Unwanted Species*. Minneapolis: University of Minnesota Press, 2013.

Nelson, Nicole. *Model Behaviour: Animal Experiments, Complexity, and the Genetics of Psychiatric Disorders*. Chicago: University of Chicago Press, 2018.

Nelson, Sara, and Bruce Braun. "Autonomia in the Anthropocene: New Challenges to Radical Politics." *South Atlantic Quarterly* 116, no. 2 (2017): 223–235.

Nelson, Sara, and Bruce Braun, eds. "Autonomia in the Anthropocene." Special issue, *South Atlantic Quarterly* 116, no. 2 (2017).

Nicholson, Marlene Arnold. "McLibel: A Case Study in English Defamation Law." *Wisconsin International Law Journal* 18 (2000): 1–145.

Nimmo, Richie. *Actor Network Theory Research*. London: Sage, 2016.

Noys, Benjamin. *The Persistence of the Negative: A Critique of Contemporary Continental Theory*. Edinburgh: Edinburgh University Press, 2012.

Nunes, Rodrigo. "Nothing Is What Democracy Looks Like." In *Shut Them Down! The G8, Gleneagles 2005 and the Movement of Movements*, edited by David Harvie, Keir Milburn, Ben Trott, and David Watts, 299–320. Leeds, UK: Dissent!; Brooklyn, NY: Autonomedia, 2005.

Oleson, Thomas. *International Zapatismo: The Construction of Solidarity in the Age of Globalization*. London: Zed Books, 2005.

Oliver, Kelly. "See Topsy 'Ride the Lightning': The Scopic Machinery of Death." *Southern Journal of Philosophy* 50, no. 1 (2012): 74–94.

Papacharissi, Zizi. "Affective Publics and Structures of Storytelling: Sentiment, Events and Mediality." *Information, Communication and Society* 19, no. 3 (2016): 307–324.

Papadopoulos, Dimitris. *Experimental Practice: Technoscience, Alterontologie, and More-than-Social Movements*. Durham, NC: Duke University Press, 2018.

Papoulias, Constantina, and Felicity Callard. "Biology's Gift: Interrogating the Turn to Affect." *Body and Society* 16, no. 1 (2010): 29–56.

Pedersen, Helena. "Release the Moths: Critical Animal Studies and the Posthumanist Impulse." *Culture, Theory and Critique* 52, no. 1 (2011): 65–81.

Philo, Chris. "Animals, Geography, and the City: Notes on Inclusions and Exclusions." *Environment and Planning D: Society and Space* 13, no. 6 (1995): 655–681.

Pick, Anat. "Executing Species: Animal Attractions in Thomas Edison and Douglas Gordon." In *The Palgrave Handbook of Posthumanism in Film and Television*, edited by Michael Hauskeller, Curtis D. Carbonell, and Thomas D. Philbeck, 311–320. Basingstoke, UK: Palgrave Macmillan, 2015.

Pickard, Victor. "Assessing the Radical Democracy of Indymedia: Discursive, Technical, and Institutional Constructions." *Critical Studies in Media Communication* 23, no. 1 (2006): 19–38.

Pickerill, Jenny. "'Autonomy Online': Indymedia and Practices of Alter-Globalisation." *Environment and Planning A: Economy and Space* 39, no. 11 (2007): 2668–2684.

Pickerill, Jenny. *Cyberprotest: Environmental Activism Online*. Manchester: Manchester University Press, 2003.

Pickerill, Jenny, and Paul Chatterton. "Notes towards Autonomous Geographies: Creation, Resistance and Self-Management as Survival Tactics." *Progress in Human Geography* 30, no. 6 (2006): 730–746.

Pickersgill, Martyn. "The Co-production of Science, Ethics, and Emotion." *Science, Technology, and Human Values* 37, no. 6 (2012): 579–603.

Pignarre, Philippe, and Isabelle Stengers. *Capitalist Sorcery: Breaking the Spell*. Translated by Andrew Goffey. London: Palgrave Macmillan, 2011.

Plumwood, Val. "Gender, Eco-feminism and the Environment." In *Controversies in Environmental Sociology*, edited by Richard White, 43–60. Cambridge: Cambridge University Press, 2004.

Plumwood, Val. "Integrating Ethical Frameworks for Animals, Humans, and Nature: A Critical Feminist Eco-socialist Analysis." *Ethics and the Environment* 5, no. 2 (2000): 285–322.

Pollock, Anne, and Banu Subramaniam, eds. "Resisting Power, Retooling Justice." Special issue, *Science, Technology, and Human Values* 14, no. 6 (2016).

Porter, Edwin S., and/or Jacob Blair Smith, dirs. *Electrocuting an Elephant*. New York: Edison Productions, 1903.

Portwood-Stacer, Laura. "Anti-consumption as Tactical Resistance: Anarchists, Subculture, and Activist Strategy." *Journal of Consumer Culture* 12, no. 1 (2012): 87–105.

Puig de la Bellacasa, Maria. *Matters of Care: Speculative Ethics in More-than-Human Worlds*. Minneapolis: University of Minnesota Press, 2017.

Puig de la Bellacasa, Maria. "Matters of Care in Technoscience: Assembling Neglected Things." *Social Studies of Science* 41, no. 1 (2011): 85–106.

Raffles, Hugh. *Insectopedia*. New York: Random House, 2011.

Rasmussen, Claire. "Pleasure, Pain and Place." In *Critical Animal Geographies: Politics, Intersections and Hierarchies in a Multispecies World*, edited by Kathryn Gillespie and Rosemary-Claire Collard, 54–70. New York: Routledge, 2015.

Ritzer, George. *The McDonaldization of Society*. London: Sage, 2013.
Ruiz, Pollyanna. *Articulating Dissent: Protest and the Public Sphere*. London: Pluto, 2014.
Sarah. "G8 on Our Doorstep." In *Shut Them Down! The G8, Gleneagles 2005 and the Movement of Movements*, edited by David Harvie, Keir Milburn, Ben Trott, and David Watts, 103–108. Leeds, UK: Dissent!; Brooklyn, NY: Autonomedia, 2005.
Sbicca, Joshua. "The Need to Feed: Urban Metabolic Struggles of Actually Existing Radical Projects." *Critical Sociology* 40, no. 6 (2013): 817–834.
Seaton, Wallis. "'Doing Her Best with What She's Got': Authorship, Irony, and Mediating Feminist Identities in Lena Dunham's *Girls*." In *Reading Lena Dunham's* Girls, edited by Elizabeth Nash and Imelda Whelehan, 149–162. Basingstoke, UK: Palgrave Macmillan, 2017.
Seaton, Wallis. "The Labour of Postfeminist Performance: Postfeminism, Authenticity and Celebrity in Representations of Girlhood on Screen." PhD diss., Keele University, 2018.
Shea, Pip, Tanya Notley, and Jean Burgess. "Editorial: Entanglements—Activism and Technology." *Fibreculture* 26 (2015): 1–6.
Shea, Pip, Tanya Notley, Jean Burgess, and Su Ballard, eds. "Entanglements—Activism and Technology." Special issue, *Fibreculture* 26 (2015).
Shotwell, Alexis. *Against Purity: Living Ethically in Compromised Times*. Minneapolis: University of Minnesota Press, 2016.
Shukin, Nicole. *Animal Capital: Rendering Life in Biopolitical Times*. Minneapolis: University of Minnesota Press, 2009.
Singer, Peter. *Animal Liberation*. London: Random House, 1995.
Sismondo, Sergio. "Post-truth?" *Social Studies of Science* 47, no. 1 (2017): 3–6.
Skeggs, Beverley, and Helen Wood, eds. *Reacting to Reality Television: Performance, Audience and Value*. Oxon, UK: Routledge, 2012.
Sorenson, John. "Constructing Terrorists: Propaganda about Animal Rights." *Critical Studies on Terrorism* 2, no. 2 (2011): 237–256.
Spurlock, Morgan, dir. *Super Size Me*. Los Angeles: Roadside Attractions, 2004.
Star, Susan Leigh. "Power, Technology and the Phenomenology of Conventions: On Being Allergic to Onions." In *A Sociology of Monsters: Essays on Power, Technology and Domination*, edited by John Law, 26–56. London: Routledge, 1991.
Star, Susan Leigh. *Regions of the Mind: Brain Research and the Quest for Scientific Certainty*. Stanford, CA: Stanford University Press, 1989.
Star, Susan Leigh. "Scientific Work and Uncertainty." *Social Studies of Science* 15, no. 3 (1985): 391–427.
Starhawk. "Diary of a Compost Toilet Queen." In *Shut Them Down! The G8, Gleneagles 2005 and the Movement of Movements*, edited by David Harvie, Keir Milburn, Ben Trott, and David Watts, 185–202. Leeds, UK: Dissent!; Brooklyn, NY: Autonomedia, 2005.
Stengers, Isabelle. "The Cosmopolitical Proposal." In *Making Things Public: Atmospheres of Democracy*, edited by Bruno Latour and Peter Weibel, 994–1003. Cambridge, MA: MIT Press, 2005.
Stengers, Isabelle. "Experimenting with Refrains: Subjectivity and the Challenge of Escaping Modern Dualism." *Subjectivity* 22, no. 1 (2008): 38–59.

Stengers, Isabelle. *In Catastrophic Times: Resisting the Coming Barbarism*. Translated by Andrew Goffey. Paris: Open Humanities Press, 2015.

Stengers, Isabelle. "Turtles All the Way Down." In *Power and Invention: Situating Science*, translated by Paul Bains, 60–74. Minneapolis: University of Minnesota Press, 1997.

Stephansen, Hilde, and Emiliano Treré. "From 'Audiences' to 'Publics': The Value of a Practice Framework for Research on Alternative and Social Movement Media." Paper presented at European Communication Research and Education Association conference, Charles University, Prague, November 10, 2016.

Stormer, Nathan. "Articulation: A Working Paper on Rhetoric and Taxis." *Quarterly Journal of Speech* 90, no. 3 (2004): 257–284.

Stringer, Tish. "This Is What Democracy Looked Like." In *Insurgent Encounters: Transnational Activism, Ethnography, and the Political*, edited by Jeffrey Juris and Alex Khasnabish, 318–341. Durham, NC: Duke University Press, 2013.

Sundberg, Juanita. "Decolonizing Posthumanist Geographies." *Cultural Geographies* 21, no. 1 (2014): 33–47.

TallBear, Kim. "Beyond the Life/Not Life Binary: A Feminist-Indigenous Reading of Cryopreservation, Interspecies Thinking, and the New Materialisms." In *Cryopolitics: Frozen Life in a Melting World*, edited by Joanna Radin and Emma Kowal, 179–200. Cambridge, MA: MIT Press, 2017.

Taylor, Sunaura. *Beasts of Burden: Animal and Disability Liberation*. New York: New Press, 2017.

Thompson, Marie. *Beyond Unwanted Sound: Noise, Affect and Aesthetic Moralism*. New York: Bloomsbury, 2017.

Thompson, Marie. "Whiteness and the Ontological Turn in Sound Studies." *Parallax* 23, no. 3 (2017): 266–282.

Thompson, Roy C. *Life-Span Effects of Radiation in the Beagle Dog*. Richland, WA: Dept. of Energy, Health and Environmental Research, Pacific Northwest Laboratory, 1989.

Todd, Zoe. "An Indigenous Feminist's Take on the Ontological Turn: 'Ontology' Is Just Another Word for Colonialism." *Journal of Historical Sociology* 29, no. 1 (2016): 4–22.

Treré, Emiliano. "Social Movements as Information Ecologies." *International Journal of Communication* 6 (2012): 2359–2377.

Trocchi, Alex, Giles Redwolf, and Petrus Alamire. "Reinventing Dissent! An Unabridged Story of Resistance." In *Shut Them Down! The G8, Gleneagles 2005 and the Movement of Movements*, edited by David Harvie, Keir Milburn, Ben Trott, and David Watts, 61–100. Leeds, UK: Dissent!; Brooklyn, NY: Autonomedia, 2005.

Tsing, Anna. *Friction: An Ethnography of Global Connection*. Princeton, NJ: Princeton University Press, 2005.

Tsing, Anna. *The Mushroom at the End of the World: On the Possibility of Life in Capitalist Ruins*. Princeton, NJ: Princeton University Press, 2015.

Tyler, Tom. *CIFERAE: A Bestiary in Five Fingers*. Minneapolis: University of Minnesota Press, 2012.

Upton, Andrew. "'Go On, Get Out There, and Make It Happen': Reflections on the First Ten Years of Stop Huntingdon Animal Cruelty (SHAC)." *Parliamentary Affairs* 65, no. 1 (2011): 238–254.

Uzelman, Scott. "Media Commons and the Sad Decline of Vancouver Indymedia." *Communication Review* 14, no. 4 (2011): 279–299.
van Dooren, Thom. "Authentic Crows: Identity, Captivity and Emergent Forms of Life." *Theory, Culture and Society* 33, no. 2 (2016): 29–52.
van Dooren, Thom. *Flight Ways: Life and Loss at the Edge of Extinction*. New York: Columbia University Press, 2014.
van Dooren, Thom, Eben Kirksey, and Ursula Münster. "Multispecies Studies: Cultivating Arts of Attentiveness." *Environmental Humanities* 8, no. 1 (2016): 1–23.
Vidal, John. *McLibel: Burger Culture on Trial*. Chatham, UK: Pan Books, 1997.
Vodovnik, Ziga, ed. *¡Ya Basta! Ten Years of the Zapatista Uprising*. Oakland, CA: AK Press, 2004.
Wadiwel, Dinesh. *The War against Animals*. Leiden: Brill, 2015.
Warin, Megan. "Foucault's Progeny: Jamie Oliver and the Art of Governing Obesity." *Social Theory and Health* 9, no. 1 (2011): 24–40.
Weisberg, Zipporah. "The Broken Promises of Monsters: Haraway, Animals and the Humanist Legacy." *Journal for Critical Animal Studies* 7, no. 2 (2009): 22–62.
Willett, Cynthia. *Interspecies Ethics*. New York: Columbia University Press, 2014.
Willey, Angela. "A World of Materialisms: Postcolonial Feminist Science Studies and the New Natural." *Science, Technology, and Human Values* 41, no. 6 (2016): 991–1014.
Wilson, Elizabeth. "Acts against Nature." *Angelaki* 23, no. 1 (2018): 19–31.
Winter, Drew Robert. "Doing Liberation: The Story and Strategy of Food Not Bombs." In *Anarchism and Animal Liberation: Essays on Complementary Elements of Total Liberation*, edited by Anthony J. Nocella II, Richard J. White, and Erika Cudworth, 59–70. Jefferson, NC: McFarland, 2015.
Wishart, Jonathan dir. *Monkeys, Rats and Me*. Aired November 27, 2006, on BBC 2.
Wolfe, Cary. *Before the Law: Humans and Other Animals in a Biopolitical Frame*. Chicago: University of Chicago Press, 2012.
Wolfe, Cary. *What Is Posthumanism?* Minneapolis: University of Minnesota Press, 2009.
Wolfson, David. *The McLibel Case and Animal Rights*. London: Active Distribution, 1999.
Wolfson, Todd. "From the Zapatistas to Indymedia: Dialectics and Orthodoxy in Contemporary Social Movements." *Communication, Culture and Critique* 5 (2012): 149–170.
Woolgar, Steve, and Javier Lezaun. "The Wrong Bin Bag: A Turn to Ontology in Science and Technology Studies?" *Social Studies of Science* 43, no. 3 (2013): 321–340.
Wrenn, Corey Lee. *A Rational Approach to Animal Rights: Extensions in Abolitionist Theory*. Basingstoke, UK: Palgrave Macmillan, 2015.
Zuiderent-Jerak, Teun, and Casper Bruun Jensen. "Editorial Introduction: Unpacking 'Intervention' in Science and Technology Studies." *Science as Culture* 16, no. 3 (2007): 227–235.
Zylinska, Joanna. *Minimal Ethics for the Anthropocene*. Ann Arbor, MI: Open Humanities Press, 2014.

Index

absent referent, visibility and, 159–167
abstraction, criticism of, 73, 202n16
activist practices: contestation of norms and, 31–35; fast-food activism and, 21–23; marginalization of, 180–182; obligations and, 6–9; personal vs. political in, 15–18; pluralisms and, 81–83; politics of openness and, 75–78; radical-participatory media experiments, 52–59, 64–68; relationality and, 3–4; risks of, 26; storytelling and, 49–51; uneven burdens of risk in, 46–68
Adams, Carol J., 159, 187nn35–36, 221n77
Adorno, Theodor, 144
advertising, legitimacy of, 41
aesthetic charisma, 125–132
affective relations: activism and, 16–18; ambivalence concerning, 142–146; animal research and, 112; charisma and, 128–132; inequalities and, 168–170; invisible work and, 36–38; sentimentality and, 144–146, 150–154
affinities, 13–15
African social media, hierarchical values and, 62–63
Against Purity (Shotwell), 163–164
agricultural chemicals, biodiversity declines and, 2, 184n4
Alaimo, Stacy, 37–38, 213n7
alternative infrastructures, cosmopolitics and, 79–81
alternative media networks, evolution of, 46–48
Anarchist Teapot collective, 84, 86–88
anarcho-punk movement, 83
Anderson, Kip, 161–162
Animal Equality group, 164–165
Animal Liberation: Devastate to Liberate or Devastatingly Liberal?, 205n50
animal rights activism: activist practices and entanglement in, 42–45, 98–117; health benefits of research and, 107–109; media portrayals of, 101–103; utilitarianism and, 132–133; “wrong way” caring and, 106–109. *See also* critical animal studies (CAS)
Anthropocene era, 184n22
anthropocentrism/anthropomorphism: animal rights activism and, 136–138; care ethics and, 138–141; embodied vulnerability and, 119–122; entanglement and, 4–7; multispecies communities and, 69–70; obligation and responsibility and, 7–9; representation and articulation and, 26–31; sentimentality and, 144–146, 150–154; visibility and, 164–167

anticapitalist practice: animal activism and, 83–85; politics of openness and, 75–78
anti-McDonald's activism: anticapitalist protest and, 83–85; evolution of, 23–26; narrative tactics in, 49–51; street campaigns and, 89–95; tactics in, 21–23
antivivisection groups: animal research and, 110–111, 208n12; media portrayals of, 101–103. *See* animal rights activism
Armstrong, Franny, 39
articulation: barriers to, 39–41; entanglement and complexity and, 42–45; invisible work of actors and, 35–38; McDonald's protests and, 25–26; representation and, 26–31; terminology relating to, 191n9
Attenborough, David (Sir), 155–156
awe, activism and logic of, 150–159
Aziz, Tipu, 106–107

Barad, Karen, 13, 46–47, 65–67, 114, 172–174, 188n49, 199n66, 200n70
Barassi, Veronica, 56, 93
Barbados Sea Turtle Project, 158
Barua, Maan, 153–154
Bastian, Michelle, 150, 155–159
Battle of Seattle (1999), 52, 59, 76, 203n27
beagles, laboratory experiments using, 118–122, 128–141, 203n24
Beasts of Burden (anonymous), 69–70, 73–74, 84, 86
Benabid, Alim Louis, 108, 110
Best, Steven, 187n35
biodiversity studies, 2, 184n4
Blue Planet (documentary series), 155–156, 217n2
Braidotti, Rosi, 185n23
Buller, Henry, 44, 187n36
burden of proof, McLibel trial and, 33–35
bureaucratic apparatus, radical participatory media and impact of, 63–64
Burgess, Jean, 64, 197n22
Butler, Judith, 190n4
Candea, Matei, 124
capacity-building practices, radical-participatory media and, 55–59
care ethics: activism and, 189n53; anthropocentrism and, 138–141; charisma and, 123–132; exclusions and, 104–105; expertise and, 112–115; hierarchies of care, 98–117; instrumentalization of, 206n23; knowledge politics and, 99–100, 103–105; rational and irrational publics and, 102–103, 208n9; uncertainty tactics and, 109–112; "wrong way" caring and, 105–109
Carnival against Capital, 59
Carpentier, Nico, 59–60
causal narrative, 106–109
celebrity promotion of environmental activism, 145–146
charisma: care and, 123–132; suffering and, 132–136
Chatterton, Paul, 51, 200n3
Chivers, Danny, 162–163
citizen camera-witnessing, 119–122
class politics: aesthetic charisma and, 217n11; anticapitalist protest and, 83–85; food activism and, 91–95
Cockburn, Alexander, 29–30
Collard, Rosemary-Claire, 10–11, 152–154
"commonsense" assertions, norms of advocacy and, 34–36
composting, protest camps and politics of, 78–81
conservation biopolitics, 72, 127–132
constraints: framing of, 22, 190n4; in McLibel trial, 22, 39–41
consumption, protest camps and, 78–81
corporeal charisma, 125–132, 213n7
"Cosmopolitical Proposal, The" (Stengers), 124
cosmopolitics: composting analogy and, 78–81; multispecies communities and, 72–75; Stengers's concept of, 50–51, 59, 71
CounterSpin Collective, 46–47, 53
Cowspiracy (documentary), 146, 161–167

critical animal studies (CAS): class politics and, 83–85; entanglement and, 4–7, 187n35; essentialism and openness and, 83–85; ethical developments in, 187n36; relational ethics and, 8–11, 73–75; visibility in, 159–167
critical-feminist theory, entanglement and, 5–7
"critique of critique" (Latour), 8, 103–106, 187n34
crittercams, 219n33
cultural theory: power relations and, 103–104; sentimentality and, 143–146
Customary Approach principle, McLibel trial and, 35
cyborg, Haraway's concept of, 149–150

Dahlgren, Peter, 59–60
"Dairy Industry in 360°, The" (documentary), 164–165
Davies, Ashley, 116
de Certeau, Michel, 18
deep brain stimulation (DBS), 108–110
Deleuze, Gilles, 143–144, 152
Della Porta, Donatella, 200n3
DeLuca, Kevin, 119
Despret, Vinciane, 124, 130–131, 165, 193n31
Dialectic of Enlightenment (Adorno & Horkheimer), 144
diffractive cultural theory, 13–15
digital media technology: activist practices and, 51; sociotechnical norms and informal hierarchies in, 59–64
Directive 2010, 214n8
Disneyfication, 143–144, 150, 212n79
disruptive tactics, infrastructural frictions and, 35–38
diversity of tactics: food activism and, 88–89; politics of openness and, 81–83
documentaries, sympathy and awe in, 151–154
dualisms, entanglements and, 183n3, 193n38
Dumbo (film), 150
ecological activism: composting and consumption and, 78–81; diffuse responsibilities and, 96–97; politics of openness and, 71–78; radical-participatory media and, 55–59
ecological charisma, 125–132
economic disparity: articulation of, 39–41, 195n67; class politics and anti-capitalist protests and, 83–85
"Eden under glass" perspective, 31–32, 156–159
Edison, Thomas, 153
Edwards, Paul N., 79
Electrocuting an Elephant (film), 152–154, 158–159
elephants, sentimentality and anthropomorphism about, 150–154
embodied care: image events and, 119–122; suffering and, 136–141; visibility and, 165–167
emotion, activism and, 16–18, 189nn52–53
encounter-based ethics, charisma and, 129–132
Entangled Empathy (Gruen), 195n21
entanglement: affinities and frictions and, 13–15; critical animal studies and, 72–75; human autonomy and, 1–3; nonhuman interactions and, 4–7; post-feminist theory and, 147–150; radical-participatory media and, 64–68; tactical interventions and, 18–20
environmental activism: entanglement and, 4–7; spectacle of, 145–146
essentialism: animal activism and, 83–85; food activism and, 88–89; politics of openness and, 71–75
exclusion ethics: care ethics and, 104–105; constitutive and creative aspects of, 172–174; entanglement and, 3–4, 184n12; future challenges for, 171–182; laboratory animal protests and, 129–132; protest movements and, 95–97; representation and, 177–180; responsibility and, 69–70; scholarship on, 9–13

Fanon, Frantz, 204n47
fast-food infrastructure: activism against, 15–18; articulation of invisible work in, 35–38; feminist theory and, 21; McLibel trial and, 31–35
Fate of the Forest, The (Hecht & Cockburn), 29–30
Feigenbaum, Anna, 64–65, 77, 203n34
Felix Campaign, 99–100, 107–108, 113, 119
feminist science studies: fast-food activism and, 21–22, 190n1; McLibel trial and, 32–35; nonhuman agency and, 192n20; popular culture and, 147–150
Fibreculture, 64, 197n22, 203n34
financial constraints, articulation and role of, 39–41
Fish, Adam, 58
Flight Ways: Life and Loss at the Edge of Extinction (van Dooren), 184n4
food activism: animal rights groups and, 98–99; articulation of invisible labor and, 35–38; fast-food infrastructure and, 15–18; feminist science and, 21; food-sharing tactics, 90–95; McLibel trial and, 31–35; prefigurative street protests and, 89–95, 204n49; in protest camps, 86–89
Food Not Bombs movement, 90–92
foreclosed realities, entanglement and, 2–3
Forsyth, Isla, 172
Foucault, Michel, 190n4
Franklin, Sarah, 206n17
Freeman, Jo, 11–12, 45, 60–61, 179, 205n63
Frenzel, Fabien, 62, 77, 203n34
frictions: definitions of, 191n16; exclusion and, 181–182; highlighting of as political tactic, 13–15, 39–41, 44–45; radical-participatory media and impact of, 63–64
Frozen Planet (documentary), 218n16

G8 Summit, 76, 78
G20 summit, 76

Gardner, John, 108
Gill, Rosalind, 147–150
Ginn, Franklin, 10–11, 45, 174
global justice movement: anticapitalist networks and, 75–78; anticapitalist protest and, 83–85; food activism and, 89–90; Haraway's characterization of, 44–45; radical-participatory media and, 52–59
global wildlife trade, entanglement ethics and, 10–11
Goodman, Mike, 145, 164–165
Graeber, David, 204n47
Green Hill beagle breeding facility, 118–122, 128–141
Greenhough, Beth, 124–125, 194n55
Gruen, Lori, 185n21
Guattari, Félix, 71, 77, 143, 152

Hall, Stuart, 191n9
Hamilton, Carrie, 189n56
Hampson, Judith, 220n64
Haraway, Donna, 5, 7, 19, 186n29; on affective relations, 142–146, 152; care ethics and, 206n23; companion species concept of, 73, 177–180; on composting and cosmopolitics, 78–81; criticism of, 188n40; on dangers of advocacy, 26–31; on disruptive tactics, 35–38, 42–45, 83, 120–122; on exclusion politics, 177–180; figurations in work of, 149–150; on global justice movement, 68; McLibel trial and theories of, 31–35; modest witnessing of, 109, 112; on multispecies interactions, 142, 214n21; on obligations and interventions, 154–159; political semiotics of representation and, 28–30; on politics of articulation, 29–31, 193nn31–32; relational ethics and, 49–51; on responsibility, 70, 202n16; "staying with the trouble" perspective of, 50, 61, 122–123, 206n23; suffering ethics criticized by, 134–136; on utilitarianism, 132–134; on veganism, 205n65, 206n17; on Zapatistas, 61

health: charisma and care in, 127–132; food activism and, 90
Hecht, Susanna, 29–30, 40, 195n72
hierarchical values: anticapitalist protest against, 83–85; care ethics and, 98–117; cosmopolitics and, 79–81; diversity of tactics and, 81–83; food activism and, 86–89, 93–95; relational ethics and, 62–64; scientific research and, 116–117; sentimentality and, 150–154
Hinchliffe, Steve, 72, 127
Hobson-West, Pru, 116
Hollin, Gregory, 127–129, 140, 172, 203n24, 222n1
Horizone activist camp, 76, 79–84, 86, 89, 93
horizontal activism, 196n3
Horkheimer, Max, 144
human exceptionalism, entanglement and, 6–9
hybridity, entanglement and, 5–7, 188n40

iAnimal initiative, 164–167
"image events," activists' use of, 118–122
Indigenous activism: scholarship on, 192n20; storytelling practices in, 50
Indymedia, 47–48, 57, 59–64, 67–68, 75–77, 92
Indymedia Mali collective, 62
inequality, activism and, 168–170
influence, activism and role of, 155–156, 217n2
informal hierarchies, 11–12; activism and, 205n63; cosmopolitics and, 79–81; exclusion and, 178–180; radical-participatory media and, 51–59; socio-technical norms and, 59–64
informed consent, embodied care and, 124–132
infrastructural frictions: cosmopolitics and, 78–81; disruptive tactics and, 35–38
International Day of Action against McDonald's, 56–57, 90
International Monetary Fund (IMF), 76
Interspecies Ethics (Willett), 185n21
intervention: entanglement and, 7–9, 184n11; posthumanism and, 185n25; relational ethics and, 223n23
invisible work, articulation of, 35–38
Irni, Sari, 192n20
issue network, 102, 209n17

Johnson, Elizabeth, 129

Kayapó people, 26–30
killability: animal activism and, 85; human exceptionalism and, 7–8, 186n29; in McLibel trial, 35
Klein, Naomi, 52
knowledge: care ethics and politics of, 100, 103–105; cosmopolitical approach to production of, 50–51; tools entangled in production of, 46–47

laboratory animals: activism on behalf of, 118–122, 203n24; care ethics and activism for, 139–141; charisma of suffering and, 132–136; nonhuman charisma and, 128–132
labor rights: articulation of, 39–41, 216n57; food activism and, 94–95; invisible work and, 35–38
Langston, William, 108–109
Latimer, Joanna, 172, 188n40
Latour, Bruno, 5, 8, 64, 103–106, 116–117
legal aid, absence of for McLibel plaintiffs, 39–41, 195n67
legislative apparatuses, political semiotics of representation and, 40–41
Leth-Epsensen, Marie, 212n79
Lezaun, Javier, 33–35
libel actions, McDonald's protests and, 23–26
Life on Earth (documentary series), 155–156
Linné, Tobias, 167
Lister, Joseph, 107
Lockwood, Alex, 161–162
London Greenpeace, 53
Lorenz, Konrad, 130–131

Lorimer, Jamie, 72, 89, 118, 125–128, 133, 144–145, 150–151, 155–156, 161–164, 202n16
Lynch, Evanna, 165
Lynn, William, 30

Madsen, Mathias Elrød, 212n79
Manhattan Project, 128–132
Mason, Peter, 208n12
McCurdy, Patrick, 77, 203n34
McGreenwash pamphlet, 57
McInformation Network, 23–26, 24
McLibel (documentary), 39
McLibel case: activists' apologies in, 191n11; barriers to articulation in, 39–41; exclusion and, 174–175; food activism and, 89–91; invisible work of actors in, 35–38; McSpotlight activism and, 52–59; media situatedness and scaling up during, 49–51; normativity and, 31–35; politics of openness and, 75–76; theoretical dilemmas posed by, 42–45. *See also* anti-McDonald's activism
McLibel Support Campaign, 24
"McLibel Two," 24–26, 35
McRobbie, Angela, 148
McSpotlight website, 21–23, 25–26, 48, 52–59
media representations of activism: accusations against activists and, 83–85, 204n46; alternative media networks, 46–48; animal rights activism and, 98–117, 209n16; care ethics and, 112–115; political semiotics of representation and, 40–41, 191n7; radical-participatory media experiments, 52–59; representation theory and, 101–103; situatedness and scale in, 48–51; sociotechnical norms and informal hierarchies in, 59–64; uncertainty tactics and, 109–112. *See also* social media
medical research, care ethics and, 108–109
memes, political semiotics of representation and, 28–30, 192n21
1-methyl-4-phenyl-1,2,3,6-tetrahydropyridine (MPTP), 108–109
Miles, Kenneth, 41
Mol, Annemarie, 41, 67, 95, 137, 174–175, 200n70
Monkeys, Rats and Me (documentary), 102, 106–115
moralism: food activism and, 93–95; openness and, 81–83
more-than-human perspective: care and charisma and, 124–132; entanglement and, 5–7; laboratory animals and, 203n24; obligation and responsibility and, 7–9, 71–75, 202n16; sentimentality and, 150–154; turtles and, 155–159
Morris, David, 23
multispecies communities: entanglement of, 69–70; exclusion ethics and, 222n1; food activism and, 88–89; laboratory animals and, 203n24; responsibility practices and, 72–75
Murphy, Michelle, 96–97

narrative of rights: causal narratives and, 106–109; framing of, 106–109; limits of, 41; McDonaldization narrative, 49–51; suffering and, 134–136
nature/culture dichotomies, political semiotics of representation and, 28–31
Nelson, Nicole, 111–112
network technologies: radical-participatory media and, 52–59, 198n40; sociotechnical norms and informal hierarchies and, 59–64
new materialism, activist tactics and, 37–38
Newsround, 217n1
No Logo (Klein), 52
nonhuman actors: charisma of, 125–132, 215n28; engagement with, 193n31, 194n55; in fast-food infrastructure, 35–38; humanist arguments and, 30–31; image events and, 119–122; scholarship on, 192n20
noninnocence: laboratory animal protests

and, 129–132; relational ethics and, 178–180, 206n23
norms and normativity: activist practices and entanglement in, 43–45; class politics and, 83–85; diversity of tactics and, 83; embodied care and, 123; food activism and, 86–89; infrastructural support and contestation for, 40–41; McDonald's protests and, 25–26; McLibel trial and, 31–35; norm congealment and, 3–4; sociotechnical norms, digital media and, 59–64
North American Free Trade Agreement (NAFTA), 61
Notley, Tanya, 64, 197n22
Nottingham Vegan Campaigns, 92
Nunes, Rodrigo, 52, 58, 77–78, 81–82, 196n3

Oncomouse, Haraway's concept of, 149–150
ontological choreography, 125–132
ontological politics, activist practices and, 38
openness, politics of: animal activism and, 83–85; anticapitalist practice and, 75–78; food activism and, 87–89; pluralisms and, 81–83; radical-participatory media and, 61–64; responsibility practices and, 71–75; risks of, 66–68
Oppose B&K Universal, 118
outreach plans, radical-participatory media and, 58, 199n48

pamphlets: animal activism and, 85; as representational advocacy, 56–57
Papadopoulos, Dimitris, 190n3, 195n65, 200n72, 223n10
Pasquali, Francesca, 60
paternalism, advocacy and, 28–30
performative tactics: food activism and, 90–95; politics of openness and, 71–75
PETA, 151, 186n30
Pickerill, Jenny, 51, 54–56, 58, 197n16, 200n3
Pignarre, Philippe, 32
Planet Earth (documentary), 155–156
Planet Earth II (documentary), 142–143, 156–159, 169
Plumwood, Val, 189n56
pluralisms, openness and, 81–83
political semiotics of representation: barriers to, 39–41; disruptive tactics and, 35–38; exclusion and, 177–180; food activism and, 92–95; nature/culture dichotomies and, 28–30; popularity and, 146–150; visibility and, 159–167
popular, politics of, 146–150, 217n14
postcolonialism: articulation research and, 192n20; representational advocacy and, 26–31
postfeminist theory, 218n21; popular culture and, 147–150
Potts, Tracey, 172
praxis: activism and, 16–18; diversity of tactics and, 81–83
primate research: activism in opposition to, 99–101; affective encounters and, 112–113; care ethics and, 109–112
"Promises of Monsters, The" (Haraway), 30, 35, 40, 43–44, 191n9
Pro-Test, 102, 111
Protest Camps (Feigenbaum, Frenzel, and McCurdy), 203n34
protest camps: diversity of tactics and, 81–83; ecological practices and, 78–81; exclusion ethics and, 95–97; food activism in, 86–89; relational ethics of, 95
proximal encounters, 100–101
public engagement, limits of, 41
public space, food activism within, 90–96
Puig de la Ballacasa, Maria: care ethics and, 98–99, 103–105, 117; class politics and, 95; on critical theory, 104–105; on entanglement and activism, 9, 42–45, 189n51; on frictions, 13; on invisible work, 36; more-than-human perspective and, 155–159; politics of opennes and, 78–79; relational ethics and, 17
"wrong way" caring and, 107–109, 1

and, 129–132; relational ethics and, 178–180, 206n23
norms and normativity: activist practices and entanglement in, 43–45; class politics and, 83–85; diversity of tactics and, 83; embodied care and, 123; food activism and, 86–89; infrastructural support and contestation for, 40–41; McDonald's protests and, 25–26; McLibel trial and, 31–35; norm congealment and, 3–4; sociotechnical norms, digital media and, 59–64
North American Free Trade Agreement (NAFTA), 61
Notley, Tanya, 64, 197n22
Nottingham Vegan Campaigns, 92
Nunes, Rodrigo, 52, 58, 77–78, 81–82, 196n3

Oncomouse, Haraway's concept of, 149–150
ontological choreography, 125–132
ontological politics, activist practices and, 38
openness, politics of: animal activism and, 83–85; anticapitalist practice and, 75–78; food activism and, 87–89; pluralisms and, 81–83; radical-participatory media and, 61–64; responsibility practices and, 71–75; risks of, 66–68
Oppose B&K Universal, 118
outreach plans, radical-participatory media and, 58, 199n48

pamphlets: animal activism and, 85; as representational advocacy, 56–57
Papadopoulos, Dimitris, 190n3, 195n65, 200n72, 223n10
Pasquali, Francesca, 60
paternalism, advocacy and, 28–30
performative tactics: food activism and, 90–95; politics of openness and, 71–75
PETA, 151, 186n30
Pickerill, Jenny, 51, 54–56, 58, 197n16, 200n3
Pignarre, Philippe, 32
Planet Earth (documentary), 155–156
Planet Earth II (documentary), 142–143, 156–159, 169
Plumwood, Val, 189n56
pluralisms, openness and, 81–83
political semiotics of representation: barriers to, 39–41; disruptive tactics and, 35–38; exclusion and, 177–180; food activism and, 92–95; nature/culture dichotomies and, 28–30; popularity and, 146–150; visibility and, 159–167
popular, politics of, 146–150, 217n14
postcolonialism: articulation research and, 192n20; representational advocacy and, 26–31
postfeminist theory, 218n21; popular culture and, 147–150
Potts, Tracey, 172
praxis: activism and, 16–18; diversity of tactics and, 81–83
primate research: activism in opposition to, 99–101; affective encounters and, 112–113; care ethics and, 109–112
"Promises of Monsters, The" (Haraway), 30, 35, 40, 43–44, 191n9
Pro-Test, 102, 111
Protest Camps (Feigenbaum, Frenzel, and McCurdy), 203n34
protest camps: diversity of tactics and, 81–83; ecological practices and, 78–81; exclusion ethics and, 95–97; food activism in, 86–89; relational ethics of, 95
proximal encounters, 100–101
public engagement, limits of, 41
public space, food activism within, 90–96
Puig de la Ballacasa, Maria: care ethics and, 98–99, 103–105, 117; class politics and, 95; on critical theory, 104–105; on entanglement and activism, 9, 42–45, 189n51; on frictions, 13; on invisible work, 36; more-than-human perspective and, 155–159; politics of openness and, 78–79; relational ethics and, 177; "wrong way" caring and, 107–109, 123

radical-participatory media experiments: activist practices and, 52–59; sociotechnical norms and informal hierarchies in, 59–64
Raffles, Hugh, 216n63
Raoni Metuktire (Chief), 26–29
Rasmussen, Claire, 160–161
Reiss, Claude, 110–111
relational ethics: affective labor and, 113–115; care ethics and, 206n23; entanglement and, 8–10, 42–45, 96–97; exclusion and, 171–182; food activism and, 21–23, 87–89; harmful life forms and, 186n31; hierarchical values and, 62–64; intervention and, 223n23; laboratory animal protests and, 129–132; nature and culture and, 186n33; politics of openness and, 76–78; popularity politics and, 146–150; suffering and, 139–141; veganism and, 206n17
remix aesthetic of digital culture, 58
representational advocacy: alternative activist media and, 46–47; ambivalence concerning, 142–146; articulation and, 26–31; laboratory animal protests and, 118–122; McLibel trial and, 31–35; media images of, 101–103; pamphlets as, 56–57; sentimentality and, 150–154, 152–154
responsibility practices: definitions of responsibility and, 200n3; exclusions and, 69–70; food activism and, 87–89, 92–95; politics of openness and, 71–75; visibility and, 159–167
rights narrative, entanglement and, 7–8
risk: activist practices and uneven burden of, 46–68; radical-participatory media and, 64–68; relational ethics and, 203n24
Roe, Emma, 124–125, 194n55
Ruiz, Pollyanna, 196n1

Safer Medicines Campaign, 102, 110, 207n5
Sbicca, Joshua, 92
scaling up of media infrastructure, 48–51
science, public opinion and, 116–117
seaborne plastic pollution, entangled responsibility for, 1–2
Seaton, Wallis, 148
self-representation: activism and, 16–18; class politics and, 83–85; nonhuman primates and, 115–117
sensibility, anthropogenesis and, 147–150
sentimentality: activism and, 142–146; care ethics and, 100–101; cultural hierarchies and, 150–154
sexuality, animal activism and role of, 159–167
Sexual Politics of Meat, The (Adams), 159
Shea, Pip, 64, 197n22
Shotwell, Alexis, 7, 44, 163–164, 184n7, 186n30
Shukin, Nicole, 152–154
Shut Them Down! (anthology), 52, 79
sick building syndrome, 96
Singer, Peter, 114, 139
situated ethics: environmental activism and, 145–146; exclusion and, 95–97; food activism and, 87–89; media infrastructure and, 48–51; relationality and, 24–26
skill sharing: composting and, 80–81; radical-participatory media and, 55–59
social media: activists use of, 191n7; alternative media networks and, 46–48; citizen camera-witnessing and, 119–122; memes in, 28–30, 192n21; mirror sites used by, 52; representational advocacy and, 54–59; sociotechnical norms and informal hierarchies in, 59–64. *See also* media representations of activism
sociotechnical norms: care ethics and, 106–109; exclusion and, 175–176; food activism in protest camps and, 86–89; informal hierarchies and, 59–64; sentimentality and, 152–154

"sorry, but we have to" logic of capitalism, 32–35, 93
SPEAK animal rights group, 98, 102, 106, 110
spectacular environmentalism, 145–146, 168–170
Spurlock, Morgan, 90
Star, Susan Leigh, 3; on invisible work, 36; norms of advocacy and, 32; politics of openness and, 75; radical-participatory media and, 64, 67; relational ethics and, 24–25, 30, 45; on sociotechnical networks, 21, 190n1; on uncertainty tactics, 110–111
Starhawk, 78–79
Staying with the Trouble (Haraway), 50, 61, 177–180
"staying with the trouble," 50, 61, 122–123, 136–138, 206n23
Steel, Helen, 23
Stengers, Isabelle, 50–51, 59, 77–79; cosmopolitics, 124–132; criticism of abstraction by, 71, 202n16; on ethical stands, 95–96; on intervention and sentimentality, 154–159; on logic of capitalism, 32, 93
Stephansen, Hilde, 55, 58
"Sticky Lives," 10–11
storytelling, ethics of, 49–51
street activism, food activism and, 89–95
structurelessness, 11–12
subjectivity of animals, primate research and, 102–103, 208n9
suffering: care ethics and, 100–101; charisma and, 132–136; embodied care and, 136–138
Sumac Centre, 89
summit protests, 76
Sundberg, Juanita, 183n3, 192n20
Super Size Me (documentary), 90
sympathy, activism and, 150–154
sympoiesis, 127

tactics: diversity of, 81–83; in food activism, 86–95; highlighting frictions as, 39–41; image events, 119–122; infrastructural frictions and, 35–38; interventions and, 18–20; in McDonald's protests, 25–26; performative tactics, 71–75; popular culture and, 146–150; theory and, 18–20; uncertainty tactics, 109–112
TallBear, Kim, 192n20
technological determinism, radical-participatory media and, 63–64
"This Image Should Be Seen by the Whole World," 26–28
Thompson, Marie, 192n20
Thousand Plateaus, A (Deleuze and Guattari), 143
Todd, Zoe, 192n20
toilets, ecological activism and role of, 79–81
Topsy the Coney Island elephant, 152–154, 169
totalizing critique, 8, 187n36, 189n56
Treré, Emiliano, 55, 58
Tsing, Anna, 48–49, 191n16
turtles, obligations and interventions and, 154–159
"Tyranny of Structurelessness, The" (Freeman), 11

uncertainty tactics, care ethics and, 109–112
Understanding Animal Research, 102
United Kingdom: animal rights activism in, 101–103; protest infrastructure in, 84–85
University of California, Davis, animal experiments at, 128–132
"untroubling trouble," 122–123
utilitarianism, charismatic suffering and, 132–133
Uzelman, Scott, 63–64

van Dooren, Thom, 49, 69, 71–74, 81, 127, 130–131, 140, 155, 173, 184n4
Veganism and Social Revolution (Dominick), 85

vegan politics, 180, 186n30, 206n17; anticapitalist protest and, 83–85; cosmopolitanism and climate change and, 202n17; food activism and, 88–89, 189n56
Veggies Catering Campaign, 56, 58, 84, 86, 89–95, 206n67
Vidal, John, 75–76
visibility, politics of, 159–167

We Have Never Been Modern (Latour), 5
Westinghouse, George, 153
What's Wrong with McDonald's? pamphlet, 56–57, 89
When Species Meet (Haraway), 44, 68, 133, 142–146, 149–150, 186n29, 205n65, 206n17
"Why Has Critique Run Out of Steam?" (Latour), 103
Wildlife in the Anthropocene (Lorimer), 118, 202n16
Willett, Cynthia, 183n21
Willey, Angela, 192n20
Wilson, Elizabeth, 173
Winter, Drew Robert, 91
Wishart, Jonathan, 107, 110
witness statements, articulation rights through, 40–41
Wolfe, Cary, 40–41, 165
Wolfson, David, 35, 61–62, 66
Woolgar, Steve, 33–34
workers' rights: articulation of, 39–41; food activism and, 94–95
World Day for Laboratory Animals, 98
World Trade Organization, 52, 59, 76
Wretched of the Earth (Fanon), 204n47

Yates-Doerr, Emily, 122–123

Zapatismo, 61–62
Zapatista National Army of Liberation, 52, 61–62
Zylinska, Joanna, 67, 200n70